Springer-Verlag   6900 Heidelberg 1 · Postfach 1780
Telefon (06221) 49101 · Telex 04-61723
1000 Berlin 33 · Heidelberger Platz 3
Telefon (0311) 822001 · Telex 01-83319

Springer-Verlag   New York, NY 10010 · 175, Fifth Avenue
New York Inc.   Telefon 673-2660

35 Fortschritte der chemischen Forschung
Topics in Current Chemistry

Inorganic Chemistry

Springer-Verlag
Berlin Heidelberg GmbH 1973

ISBN 978-3-540-06080-2   ISBN 978-3-540-38067-2 (eBook)
DOI 10.1007/978-3-540-38067-2

Library of Congress Cata-log Card Number 51-5497.

Contents

The Chemistry of Phosphine

Prof. Dr. Ekkehard Fluck

Institut für Anorganische Chemie der Universität Stuttgart*

Contents

---

* The present article is a revised and extended version of the work previously published by Fluck and Novobilsky: Fortschr. chem. Forsch. *13*, 125 (1969). It is especially concerned with the many new physico-chemical investigations which have been carried out on phosphine. The literature has been covered to the end of 1971 and in some cases beyond this date.

1

I. Introduction

Since the discovery of phosphine by Gengembre and Kirwan [1,2,3] , the first published reports about its formation [264, 265] and the relatively early investigations of its reactions with heated alkali metals [4,5] the compound has been mentioned in all text-books and compilations. The study of its chemical behaviour was, however, only carried out systematically in the last twenty years. Nuclear magnetic resonance experiments on phosphine and its inorganic derivatives and the attempted correlation of the data with the chemical properties of the compounds induced us to review the literature on phosphine and its reactions. The results of this work are presented in the following article.

II. Properties of Phosphine and its Determination

1. Physical Properties and Data

At room temperature phosphine is a *colourless gas* with a garliclike smell. Below $-87.74\,°C$ [6] [$-87.44\,°C$ [7]; $-87.78\,°C$ [8]; $-87.9\,°C$ [341]], it is a colourless liquid which freezes at $-133.5\,°C$ [7,8,9]. Solid phosphorus hydride exists in four or five different forms [7,8,10]. The transformations occuring at $-185, -224, -243$ and $-263\,°C$ are possibly connected with the hindered rotation of the molecule. The weight of phosphine at normal temperature and pressure is $1.5307\ g/l$ [11,12], the density of liquid phosphine at $-90\,°C$ is $0.746\ g/cm^3$ [11,12,321]. The density of liquid phosphine in the temperature range between the triple-point and the boiling point can be obtained from the equation

$$d_4^t = 0.594 - (1.71 \cdot 10^{-3} t \pm 3 \cdot 10^{-5} t) \tag{1}$$

where d_4^t is the density of the hydride relative to the density of water at $4\,°C$ and t is the temperature [342]. The density of solid phosphine was found to be $0.896\ g/cm^3$ at $-135\,°C$ [13]. The vapour pressure over liquid phosphine below the boiling point can be determined by the equation [8,14,284]

$$\log p\ (cm \cdot Hg) = -1027.300/T - 0.0178530\ T + 0.000029135\ T^2 + 9.73075 \tag{2}$$

Vapour pressures of phosphine at temperatures $< 25\,°C$ were determined by A. Stock *et al.* [15] and Stephenson and Giauque [8], and between 25 and $50\,°C$ by Briner [269]. See also [270,271]. The vapour pressures of the system PH_3/AsH_3 in the temperature range -78.7 to $-100\,°C$ were determined by Devyatykh *et al.* [341].

According to the most recent measurements the critical pressure is 65 at and the critical temperature $52\,°C$ [272]. Earlier work gave similar values [16,269, 273,281]. The latent heat of vapourisation of phosphine at the boiling point was found experimentally to be 3493 ± 3 cal/mol [6,17], a more recent value is 3949 cal/mol [341]. The molar entropy at the boiling point is 46.93 cal·grad^{-1} [7,8]. Earlier work gave the remarkably small value of 18.8 for the Trouton constant [7]. However, the latest report of 21.3 is probably correct [341].

The equilibria between the liquid and vapour of binary mixtures of phosphine and arsine were investigated by Devyatykh et al. [341].

The surface tension of liquid phosphine at –100 °C is 22.0 dyn/cm [319, 320,349].

The viscosity of phosphine at 273 K is 1073 · 10^{-7} poise [343,344]. The temperature dependence can be represented by the equation

$$\eta = K \cdot T^S \tag{3}$$

where η is the viscosity coefficient (g · cm^{-1} · s^{-1}), T the temperature in Kelvin, K and S are constants. For phosphine in the temperature range 193 – 273 K, $K = 3.648 \cdot 10^{-7}$ and $S = 1.013$ [343].

Phosphine is thermally very stable and decomposes noticeably only above 550 °C [274,275].

At –140 °C *solid phosphine* crystallises in a face-centred cubic form with four molecules in the unit cell [13]. The lattice constant is $a = 6.31 \pm 0.01$ Å [13]. Agreeing results from various different methods show that the phosphine molecule has a pyramidal structure with C_{3v} symmetry [18-24]. Also infra-red and Raman [25,26], micro-wave [27-29], and nuclear magnetic resonance experiments [30] confirm this result, as do the findings from electron diffraction experiments [31-33]. Helminger and Gordy [345] recently carried out thorough investigations of the micro-wave spectra (sub-millimetre wave spectra) of phosphine and deuterophosphine. Infra-red experiments suggest that the phosphine molecule in the solid state also has C_{3v} or C_3 symmetry [10,13,328]. X-Ray diffraction studies on solid phosphine have not yet been carried out.

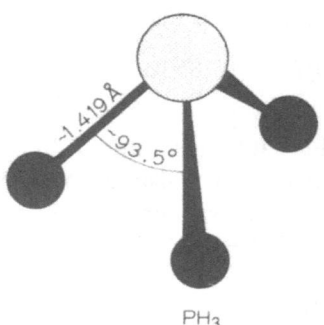

PH₃

Fig. 1. PH$_3$ molecule (bond angle and interatomic distance)

The values obtained for the P–H bond lengths and H–P–H bond angles by different methods are collected together in Table 1. Calculations on the bond lengths in PH_3, PH_4^+ and PH_2^- have been made by Banyard and Hake [41]. *Ab initio* SCF–LCAO–MO calculations for the phosphine molecule have been made recently by various workers and the results compared with the experimentally determined physical quantities for the molecule [346,347].

Revisions of earlier data on the *heat of formation* and the bond energies of phosphine have been undertaken [40]. The heat of formation ΔH_{f_0} of phosphine from white phosphorus and hydrogen is 1.30 kcal/mol [40], (the corresponding value for P_2H_4 is 5.0 ± 1.0 kcal/mol), the average bond energy E (P–H) was found to be 76.8 kcal.

When subjected to shock-waves, phosphine dissociates into hydrogen and red phosphorus. The radiation thus emitted is only visible in reflection. In contrast, the shock-wave induced dissociation of phosphine diluted with argon is accompanied by emission of visible light. In this case the reaction products are hydrogen and white phosphorus [465].

The refractive index between gaseous phosphine at 0 °C and 760 mmHg and vacuum is 1.000789 when measured with white light [466]. The refractive indices between liquid phosphine and air were found to be, for white light at 11 °C, 1.323 and for the Na–D-line at 17.5 °C, 1.317 [467,468].

Measurements of the magnetic susceptibility at room temperature gave a value of $\chi = -26.2 \cdot 10^{-6}$ cm^3/g [469]. The magnetic rotation (Faraday Effect) of gasous phosphine at 0 °C, 1 atm and $\lambda = 5700$ Å was measured as $[M]_\omega = 57 \cdot 10^{-6}$ rad m^4 V^{-1} s^{-1} = 0.30 min cm^2 Gauss^{-1} mol^{-1} [470].

Phosphine has a *dipole moment* of 0.58 D [42,43]; on substituting all of the hydrogen atoms by methyl or ethyl groups it increases to 1.19 or 1.35 D, respectively [43]. In the series of compounds PH_3 to $P(CH_3)_3$ the following values for the dipole moments were measured: PH_3 0.579 D, CH_3PH_2 1.100 D, $(CH_3)_2PH$ 1.230 D, $(CH_3)_3P$ 1.192 D. These results are in close agreement with the figures calculated on the assumption that the molecules have pyramidal structures. The most probable values for the individual bond moments are m_0 (P–H) = 0.371 D and m_0 (P–C) = 0.690 D [348]. The dipole moment of triethyl-phosphine is 1.35 D [43]. In contrast to phosphine, the dipole moment of ammonia (1.47 D [43]) decreases when the hydrogen atoms are replaced by alkyl groups. Both trimethyl- and triethylamine have dipole moments of 0.61 D [43]. Weaver and Parry [43] interpreted this result as follows: the contribution of the electron lone pair to the dipole moment of phosphine is very small whereas that of the electron lone pair in ammonia is very large (see also [44,45]). This assumption is confirmed by the chemical behaviour and nuclear magnetic resonance spectra of phosphine, which also suggest that practically pure *p* orbitals of phosphorus are used for the P–H bonds [46-48].

The ^1H *nuclear magnetic resonance spectrum* of phosphine dissolved in liquid ammonia shows a chemical shift of $\delta_H = +1.66$ ppm (relative to $(CH_3)_4Si$)

Table 1. Bond lengths and bond angles in the phosphorus hydride molecule and some of its derivatives (from Corbridge [318])

Compound	Bond length [Å] P–H or P–D	Angle [°] H–P–H	Method	Lit.
PH_3	1.415 ± 0.005		ED/MI[a]	[34]
	1.44		calc.	[35]
	1.437 ± 0.004		ED	[33]
	1.4206 ± 0.005	93.5	MI	[27]
	1.419	93.8	IR[a]	[20]
	1.419	93.5	MI	[28]
	1.418	93.3	calc.	[36]
CH_3PH_2	1.414 ± 0.003	93.4	MI	[37]
	1.445 ± 0.007	–	ED	[31]
$(CH_3)_2PH$	1.445 ± 0.02	–	ED	[31]
PHD_2	1.4116 ± 0.0001	93.2	MI	[29]
PH_2D	1.4177 ± 0.0001	93.4	MI	[29]
PD_3	1.4166 ± 0.005	–	MI	[27]
PH_4I	1.42 ± 0.02	–	NMR[a]	[38]
PH	1.43		IR	[39]

[a] ED: Electron diffraction; MI: Micro-wave; IR: Infra-red; NMR: Nuclear magnetic resonance.

and a coupling constant of J_{H-P} = 188.7 Hz. The coupling constants in the pure liquid were reported to be J_{H-P} = 182.2 ± 0.3 Hz and J_{H-H} = 13.2 ± 0.7 Hz [49]. Ebsworth and Sheldrick [145] studied the dependence of the chemical shift and H–P coupling constant of phosphine on concentration, temperature and solvent. Two phases are formed in fairly concentrated solutions of phosphine in liquid ammonia below –30 °C, one of these is phosphine-rich and the other ammonia-rich. In the phosphine-rich phase, the coupling constant, J_{H-P}, increases from 185.2 to 186.6 Hz on cooling from –32 to –79 °C, while in the ammonia-rich phase between the same temperatures it increases from 191.1 to 195.1 Hz.

The difference between the chemical shift, δ_H, for gaseous and for liquid phosphine is small [266]; this indicates that the degree of association in the liquid phase is small. Consideration of thermal data [267,268] leads to the same conclusion. In comparison to water or ammonia, phosphine forms practically no hydrogen bonds.

Birchall and Jolly used ^1H NMR data for phosphine, arsine, and germane and some of their alkyl derivatives to determine the relative acidities in liquid ammonia [259]. Spin-lattice relaxation time (T_1) measurements for the ^1H nuclei in PH_3 are reported by Armstrong and Courtney [350].

The expected $1 : 3 : 3 : 1$ quartet is observed in the ^{31}P *nuclear magnetic resonance spectrum*. The chemical shift, δ_P, relative to 85% aqueous orthophosphoric acid, is $+241.0$ ppm. The high positive chemical shift of the phosphorus atom is clearly related to the fact that the bonds between the central phosphorus atom and the hydrogen atoms are almost pure $p_\sigma(P) - s_\sigma(H)$ bonds. This is in accord with the observed bond angle of *ca.* 93°, numerous physical data [46-48], and theoretical considerations [50-52]. Calculations of the overlap integrals of the $s - p$ functions of the phosphorus atom with the hydrogen $1s$ function show that the $3s(P)-1s(H)$ overlap integral is smaller than the $3p_\sigma(P)-1s(H)$ overlap integral, so that the small deviation of the bond angle from 90° probably results from mutual repulsion of the hydrogen atoms [52] (see also [338]). The two free electrons in the valence shell remain in an s orbital and so are relatively near to the nucleus. This results in a strong shielding of the phosphorus nucleus which, in turn, causes the high chemical shift. This particular electronic configuration is also apparent from the chemical behaviour of the molecule; the nucleophilic character is not very pronounced so that phosphine is very unreactive in many respects. This inertness is particularly noticeable when the chemical behaviour of phosphine is compared with that of the alkyl-phosphines. In the latter, the σ-bond system contains a large contribution from the $3s$ orbital so that the electron lone pairs have considerable p character. In comparison to those in the phosphine molecule, these are in much more wide-reaching orbitals and can therefore take part more easily in the first step of a nucleophilic attack.

The *infra-red spectra* of gaseous phosphine have been described and interpreted by different authors [19,25,53-56], that of solid phosphine has been reported by Heinemann [328]. The infra-red spectra of solid phosphine in the temperature range 4 to 68 K have been measured by Hardin and Harvey [10]. At 10 K a previously undescribed transformation was observed.

The molecular data obtained from the infra-red spectra (see Table 1) are in good agreement with the results from *micro-wave spectra* [28,29].

The gas phase *UV-absorption spectra* of phosphine and a series of other phosphorus compounds have been studied in the long wave-length region [57-59] and recently by Halmann [60] in the $1850 - 2500$ Å range. In general, it is assumed that the electronic excitation process of lowest energy in compounds, such as NH_3, H_2O, PH_3, H_2S or HCl and their alkyl derivatives, is that which involves the promotion of non-bonding electrons of the most electronegative atom into an anti-bonding orbital, *i.e.* an $n \rightarrow \sigma^*$ transition. However, Halmann [60] attributes a strong absorption band in the spectrum of phosphine at 1910 Å to an $n \rightarrow \delta^*$ transition. The wave lengths at which the maximum absorptions occur and their extinction coefficients are summarised in Table 2 and compared with the corresponding values for ammonia and amines [57,61].

The absorption of phosphine at even shorter wave-lengths (down to 1250 Å) was measured by Walsh and Warsop [62]. The ionisation potentials of phosphine were found, from the 584 Å photoelectron spectrum, to be 10.28 and 12.90 eV [351]. See also [352].

Comprehensive investigations of the *mass spectra* of phosphine and diphosphine were carried out by Wada and Kiser [63], among others [262,353]. The ionisation potential of phosphine was found to be 10.2 ± 0.2 eV, in good agreement with the experiments of Neuert and Clasen [64]. Deviating values were reported by Saalfeld and Svec [65,66]. The ionisation potential of diphosphine is 8.7 ± 0.3 eV. The corresponding ionisation potentials for ammonia and hydrazine are 10.15 – 10.5 [67-69] and 9.00 ± 0.1 eV [70], respectively. The $P_2H_3^+$ ions

Table 2. UV data for phosphines and amines [60]

Compound	λ_{max} [Å]	$\epsilon[l \cdot mol^{-1} \cdot cm^{-1}]$
PH_3	1910	3400 ± 200
CH_3PH_2	2010 ± 3	130 ± 30
	1960	weak
	1870	1500 ± 200
$(CH_3)_2PH$	1890	6300 ± 500
$(CH_3)_3P$	2010 ± 5	18800 ± 100
NH_3	1942	5600
	1515	strong
CH_3NH_2	2150	600
	1737	2200
$(CH_3)_2NH$	2200	100
	1905	3300
$(CH_3)_3N$	2273	900
	1990	3950

observed in the mass spectrum of diphosphine do not, according to Wada and Kiser, arise from the reaction

$$e + P_2H_4 \longrightarrow P_2H_3^+ + H + 2e \tag{4}$$

but from the simple ionisation of P_2H_3 radicals which are formed as intermediates in the thermal decomposition of diphosphine to PH_3 and a solid reported to have the approximate constitution P_2H. The appearance of the PH_4^+ ion in the mass spectra of phosphine results from the reaction [262]

$$PH_3^+ + PH_3 \longrightarrow PH_4^+ + PH_2 \tag{5}$$

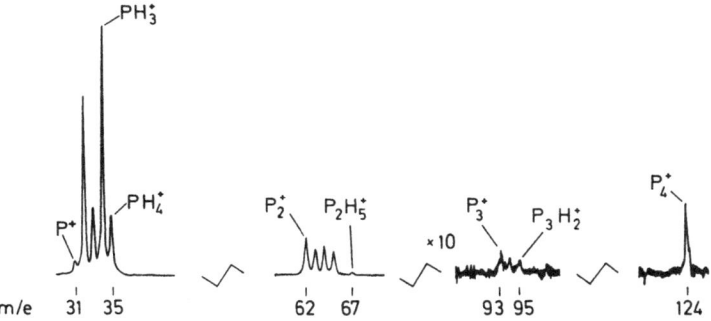

Fig. 2. ICR spectrum of the positive ions of PH_3 at $2 \cdot 10^{-5}$ Torr and 23 eV ionisation energy

Further ion-molecule reactions were identified in the gas phase by Eyler [354] using ion cyclotron single and double resonance. The ion cyclotron single resonance spectrum of phosphine is shown in Fig. 2.
As well as the PH_3^+ ion, the ions PH_2^+, PH^+ and P^+, caused by fragmentation are observed. Furthermore the signal for the phosphonium ion, formed according to Eq. (5), is seen. Many other, heavier ions are the products of ion molecule reactions in PH_3. These have the general formulae $P_2H_n^+$ ($n = 0 - 5$), $P_3H_n^+$ ($n = 0 - 2$), and P_4^+. Analogous ions were also formed by ion molecule reactions in ammonia. The reactions listed in Table 3 were identified with the help of the ion cyclotron double resonance technique.
The results of elctron-impact studies of phosphine by Halmann et al. [451] are given in Table 3a. The authors used the appearance potentials, in conjunction with thermochemical data, to choose the probable reaction processes. In many simple cases the observed appearance potential $A(Z)$ for an ion fragment Z from a molecule RZ is related to its ionisation potential $I(Z)$ and to the energy of dissociation $D(R-Z)$ of the bond by the expression $A(Z) = I(Z) + D(R-Z)$. This assumes that the dissociation products are formed with little, if any, excitation energy, and that $I(Z) < I(R)$. The most abundant ion species in the usual mass spectrum of phosphine is PH^+, which is probably formed according to the following mechanism

$$PH_3 + e \rightarrow PH^+ + H_2 + 2e$$

The appearance potential for this reaction should be

$$I(PH) \leqslant A(PH^+) - 2\overline{D}(P-H) + D(H-H) = 13.1 - 6.7 + 4.48 = 10.9 \pm 0.5 \text{ eV}.$$

The average energy of dissociation of the P–H-bond is known from thermochemical measurements, $\overline{D}(P-H) = 3.35$ eV. The dissociation energy of the hydrogen molecule is $D(H-H) = 4.48$ eV. The appearance potential for PH^+ formed according to the mechanism

9

Table 3. Ion molecule reactions of phosphine
(obtained from ion cyclotron double resonance,
the neutral species are assumed) [354]

PH_3^+ + PH_3	$\longrightarrow$	PH_4^+ + PH_2	
PH_3^+ + PH_3	$\longrightarrow$	P_2H^+ + $2H_2$ + H	
PH_3^+ + PH_3	$\longrightarrow$	$P_2H_3^+$ + H_2 + H	
PH_3^+ + PH_3	$\longrightarrow$	$P_2H_4^+$ + H_2	
PH_3^+ + PH_3	$\longrightarrow$	$P_2H_5^+$ + H	
PH_2^+ + PH_3	$\longrightarrow$	PH_3^+ + PH_2	
PH_2^+ + PH_3	$\longrightarrow$	P_2H^+ + $2H_2$	
PH_2^+ + PH_3	$\longrightarrow$	$P_2H_3^+$ + H_2	
PH^+ + PH_3	$\longrightarrow$	PH_4^+ + P	
PH^+ + PH_3	$\longrightarrow$	PH_3^+ + PH	
PH^+ + PH_3	$\longrightarrow$	PH_2^+ + PH_2	
PH^+ + PH_3	$\longrightarrow$	P_2^+ + $2H_2$	
PH^+ + PH_3	$\longrightarrow$	$P_2H_2^+$ + H_2	
PH^+ + PH_3	$\longrightarrow$	$P_2H_3^+$ + H	
P^+ + PH_3	$\longrightarrow$	PH_3^+ + P	
P^+ + PH_3	$\longrightarrow$	P_2H^+ + H_2	
P_2^+ + PH_3	$\longrightarrow$	P_3H^+ + H_2	
P_2^+ + PH_3	$\longrightarrow$	$P_3H_2^+$ + H	
P_2H^+ + PH_3	$\longrightarrow$	$P_3H_2^+$ + H_2	

$$PH_3 + e \rightarrow PH^+ + 2H + 2e$$

would be higher by the amount of energy of dissociation of the hydrogen
molecule.

The formation of P^+ by electron impact on phosphine may be due to the
following process:

$$PH_3 + e \rightarrow P^+ + H_2 + H + 2e$$

The lowest appearance potential for P^+ can be predicted by the equation
$A(P^+) \geqslant I(P) + 3\bar{D}(P-H) - D(H-H)$. The ionisation potential of the phosphorus atom has been determined spectroscopically, namely $I(P) = 11.0$ eV.
Thus $A(P^+) = 11.0 + 3 \cdot 3.35 - 4.48 = 16.6$ eV. This predicted value for the
process mentioned above is close to the observed value for the "vanishing
current" appearance potential of P^+. The "linear extrapolation" value of
20 eV may be due to a process in which three hydrogen atoms are formed.

For the formation of doubly charged phosphorus ions, P^{2+}, the following
mechanism is suggested

$$PH_3 + e \longrightarrow P^{2+} + 3H + 3e$$

The appearance potential for this ion should thus be equal to, or larger than, the sum of the potentials for double ionisation of phosphorus, $I(P^I) + I(P^{II})$, and the dissociation energy of phosphine should be given by

$$A(P^{2+}) \geqslant I(P^I) + I(P^{II}) + 3\bar{D}(P-H)$$

With the spectroscopic values for $I(P^I) = 11.0$ eV and $I(P^{II}) = 19.65$ eV the predicted value of $A(P^{2+})$ is 40.7 eV. The agreement between this and the observed value of 42 eV seems to support the suggested dissociation mechanism

Table 3a. Appearance potentials A for ion fragments from phosphine [451]

Ion	m/e	Reference ion	A [eV] linear extrapol.	A [eV] vanishing current
PH_3^+	34	Ar^+	10.3 ± 0.5	10.4 ± 0.3
PH_2^+	33	Ar^+	14.4	14.0 ± 0.2
PH^+	32	Ar^+	13.6	13.1 ± 0.2
P^+	31	Ar^+	20 ± 1	16.0 ± 1.0
			Linear extrapol.	Square root plot
PH_3^{2+}	17	Kr^{2+}	15.0	15.6
PH_2^{2+}	16.5	Kr^{2+}	32.7	34.0
PH^{2+}	16	Kr^{2+}	21.2	15.1
P^{2+}	15.5	Kr^{2+}	42 ± 2	42 ± 2

The flash photolysis of phosphine, according to spectroscopic results, causes the formation of two phosphorus- and two hydrogen-containing radicals, corresponding to the dissociation of phosphine as shown in Eqs. (6) and (7) [71].

$$PH_3 \xrightarrow{h \cdot \nu} PH_2 + H \tag{6}$$

$$PH_3 \xrightarrow{h \cdot \nu} PH + H_2 \tag{7}$$

In addition, Basco and Yee observed electronically excited phosphorus atoms and P_2 molecules in excited vibration states in the absorption spectra [355]. The latter were also found by Norrish and Oldershaw [72].

Phosphine, subjected to radiation of the 10.59 μ-line of a CO_2 laser dissociates into the elements [261].

Neutron irradiation of gaseous phosphine always results in 40 to 60 % of the ^{32}P being retained as ^{32}PH$_3$ while the balance is deposited on the walls as phosphorus oxyacids [452-454]. This result is independent of whether the irradiation is done in the presence of an excess of various inert gases or methane (as possible moderators for hot atoms), or in the presence of substances which could possibly scavenge thermal phosphorus atoms. This indicates that "hot" phosphorus atoms do not form stable products and that phosphine itself is a very efficient scavenger for thermal phosphorus atoms. When phosphine was irradiated in a large excess of methane Halmann and Kugel [452] observed the formation of ^{32}P-labelled methyl-, dimethyl- and trimethyl-phosphines. The addition of traces of water enhanced the yields of methylphosphines. As a possible reactive intermediate Halmann and Kugel suggest H–C≡P, which may account for products such as CH_3PH_2, $CH_3P(H)(O)(OH)$ and $CH_3PO_3H_2$, while $(CH_2)_2P$ may be a hypothetical intermediate for dimethylphosphine and its oxidation products. Also the thermal-neutron irradiation of trimethylphosphine results in the formation of phosphine, in which, among the volatile products, most of the radioactivity is found (PH$_3$ 40%, CH_3PH_2 1–3%, $(CH_3)_2PH$ 0.3 %, $(CH_3)_3P$ 0.6 – 3.8 % of the total activity). The activity in trimethylphosphine decreased with lower pressure. This demonstrates that there is no appreciable retention of chemical bonds of the recoiling phosphorus atoms [455].

The solubility of phosphine in water is, in comparison to ammonia, very small [73,74]. At 17 °C, only 22.8 ml of gaseous phosphine dissolve in 100 ml of water [73].

The Ostwald solubility coefficient β, i.e. the ratio of the concentration of the dissolved substance in the solution phase to the concentration of the dissolved substance in the gas phase is, at 297.5 K, 0.201. In the pressure range of 100 to 700 mmHg it is independent of pressure so that, at least below 1 atm, Henry's law is obeyed. With increasing temperature β decreases and reaches a value of 0.137 at 323.2 K. The enthalpy of solution calculated from the temperature dependence of β is –2.95 ± 0.1 kcal/mol [471].

According to earlier results, the solubilities at 18 °C in ethanol, ether and oil of turpentine are 0.5, 2 and 3.25 volumes of phosphine per volume of solvent, respectively [357]. Phosphine dissolves in cyclohexanol far more readily than in water. At 26 °C and a partial pressure of 766 mmHg, 2856 ml of phosphine dissolve in 1000 ml of cyclohexanol [322]. 15,900 ml of phosphine dissolve in 1 litre of trifluoroacetic acid at 26 °C and a pressure of 653 mmHg [358]. For the solubilities of phosphine in non-polar solvents see [312,315].

When phosphine is liquified by pressure in the presence of water, it dissolves partly in the water, the rest floats on the solution. If the pressure is suddenly removed colourless crystals of phosphine hydrate are formed at 2.20 °C under a pressure of 2.8 atm, 11 °C under 6.7 atm, and at 20.0 °C under 151.1 atm. When the pressure is reduced too rapidly the crystals dissolve again, also they do not form at temperatures above 28 °C. In the presence of CO_2 crystals are formed which are stable at 22 °C. In the presence of phosphine and water, CS_2 behaves similarly to CO_2 [472,473]. In all cases clathrate compounds are formed. The cubic unit cell of phosphine hydrate contains 46 molecules of water. Their skeleton forms 2 cavities with a coordination number of 20 and 6 with a coordination number of 24. If all the cavities were occupied by phosphine molecules a composition of $PH_3 \cdot 5.75\ H_2O$ or $8\ PH_3 \cdot 46\ H_2O$ would be obtained. In practice, crystals with a composition of $PH_3 \cdot 5.9\ H_2O$ are found in the above described experiments. The dissociation pressure at 0 °C is 1.6 atm, the decomposition temperature at 1 atm is −6.4 °C and the critical decomposition temperature is 28 °C [474,475,476]

Aqueous solutions of phosphine show neither acidic nor basic properties. Weston and Bigeleisen [74] investigated the deuterium exchange between D_2O and PH_3. It was found that this exchange proceeds *via* a PH_4^+ ion in acidic solutions and *via* a PH_2^- ion in basic solutions. From the kinetic data and the assumed exchange mechanisms, these authors calculated the equilibrium constants at 27 °C for reactions (8) and (9) to be $K_b \approx 4 \cdot 10^{-28}$ and $K_a \approx 1.6 \cdot 10^{-29}$, respectively.

$$PH_3 + H_2O \rightleftharpoons PH_4^+ + OH^- \tag{8}$$

$$PH_3 + H_2O \rightleftharpoons PH_2^- + H_3O^+ \tag{9}$$

These small constants suggest that the acidic or basic properties of phosphine can only be observed under special circumstances. For example, phosphine behaves as a base when it is dissolved in very strong acids. In concentrated sulphuric acid, $BF_3 \cdot H_2O$ (with excess BF_3) or in $BF_3 \cdot CH_3OH$ (with excess BF_3), phosphine accepts a proton to form a phosphonium ion, which could be identified in solution for the first time by proton and phosphorus nuclear magnetic resonance spectroscopy. The proton spectrum ascribed to the PH_4^+ ion in sulphuric acid consists of a 1 : 1 doublet with a coupling constant of approximately 547 Hz. Similar doublets are also observed in the spectra of PH_3 in the other solvents mentioned. The ^{31}P spectra of solutions of phosphine in $BF_3 \cdot H_2O$ and $BF_3 \cdot CH_3OH$, which are stable at room temperature, show 1 : 4 : 6 : 4 : 1 quintets. These are conclusive proof that PH_4^+ ions exist in these solutions. The chemical shift δ_P is +217 ppm (relative to P_4O_6), while τ was found to be 3.84 [PH_3 in 98% H_2SO_4 + $(CH_3)_4PCl$ at −43 °C] [75].

Also fluorosulphonic acid protonates phosphine as well as organophosphines [356]. The phosphonium ions formed are soluble in fluorosulphonic acid. The chemical shifts, δ_{31P} and δ_{1H}, of phosphine, the phosphonium ion and a series of organophosphines and the respective cations obtained by protonation are shown in Table 4.

Banyard and Hake [41] calculated the molecular energies for PH_3, PH_4^+ and PH_2^-. From the difference between the energies for PH_3 and PH_4^+ the proton

Table 4. N.M.R. data for phosphines and phosphonium ions [356]

	Chemical shift[a] δ_P [ppm]	Chemical shift δ_H [ppm]	J_{H-P} [Hz]
Phosphine			
PH_3	$+238$	2.28 (in CCl_4)	188
$P(CH_3)_3$	$+\ 62.2$		
$P(C_2H_5)_3$	$+\ 19.2$		
$P(i-C_3H_7)_3$	$-\ 19.3$		
$P(t-C_4H_9)_3$	$-\ 61.9$		
$P(n-C_4H_9)_3$	$+\ 32.6$		
$P(n-C_8H_{17})_3$	$+\ 32.5$		
$P(c-C_6H_{11})_3$	$-\ 11.3$ (in $CHCl_3$)		
$P(C_6H_5)_3$	$+\ \ 5.4$ (in CCl_4)		
$HP(C_6H_5)_2$	$+\ 40.7$		
Phosphonium ions			
PH_4^+	$+101.0$	6.20	548
$[HP(CH_3)_3]^+$	$+\ \ 3.2$	6.36	497
$[HP(C_2H_5)_3]^+$	$-\ 22.5$	5.97	471
$[HP(i-C_3H_7)_3]^+$	$-\ 44.4$	5.58	448
$[HP(t-C_4H_9)_3]^+$	$-\ 58.3$	5.46	436
$[HP(n-C_4H_9)_3]^+$	$-\ 13.7$	6.01	470
$[HP(n-C_8H_{17})_3]^+$	$-\ 13.0$	6.04	465
$[HP(c-C_6H_{11})_3]^+$	$-\ 32.7$	5.48	445
$[HP(C_6H_5)_3]^+$	$-\ \ 6.8$	8.48	510
$[H_2P(C_6H_5)_2]^+$	$+\ 21.2$	7.88	519

[a] The chemical shifts, δ_{31P} are all relative to 85% aqueous orthophosphoric acid.

affinity of phosphine is found to be 236 kcal/mol. This value is, however, too high. A lower figure was obtained by Wendlandt [76] using a Born-Haber cycle:

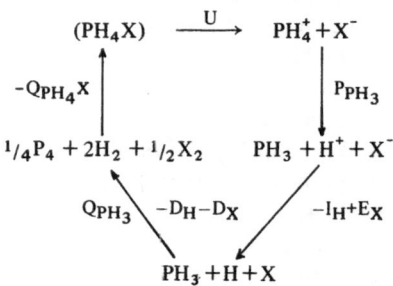

The proton affinity, P_{PH_3}, at 0 K is given by:

$$P_{PH_3} = U + Q_{PH_4X} - Q_{PH_3} + D_H + I_H + D_X - E_X - 5/2\,RT \qquad (10)$$

where U is the lattice energy of PH_4X, Q_{PH_4X} the heat of formation of PH_4X, Q_{PH_3} the heat of formation of PH_3, D_H the heat of dissociation of hydrogen, I_H the ionisation potential of hydrogen, D_X the heat of dissociation of the halogen molecule, E_X the electron affinity of the halogen, and R the gas constant $(1.987 \text{ cal·grad}^{-1} \cdot \text{mol}^{-1})$. Using the values for phosphonium iodide, PH_4I, Eq. (10) gives the proton affinity of phosphine as 200 ± 10 kcal/mol.

U(CsCl lattice)	131.5 kcal/mol
$-Q_{PH_4X}$	15.8
Q_{PH_3}	2.21
$-D_H$	52.1
$-I_H$	311.9
$-D_X$	25.5
E_X	74.6
$5/2\,RT$	1.5

Waddington [77] reported an approximately similar value of 194.5 ± 5 kcal/mol. Both these results agree well with that reported by Holtz and Beauchamp [359] for the proton affinity of phosphine. These workers determined the proton affinity from ion molecule reactions of the type

$$M_1 + M_2H^+ \rightleftharpoons M_1H^+ + M_2 \qquad (11)$$

which take place with negligible energies of activation. Such a reaction proceeds to the right only when the proton affinity of M_1 is greater than or equal to that of M_2. Binary mixtures of phosphine with acetaldehyde, acetone, ammonia or water were investigated at pressures in the range of 10^{-7} to 10^{-4} mmHg using ion cyclotron resonance. At pressures $> 10^{-6}$ mmHg, proton exchanges as shown by the general Eq. (11) were observed. The results of these investigations are summarised in Table 5.

Table 5. Protonation reactions

No.	Observed reaction	Proton affinity of PH$_3$ [kcal/mol]	Lit.
1	$PH_4^+ + NH_3 \rightarrow NH_4^+ + PH_3$	207	362)
2	$PH_4^+ + (CH_3)_2CO \rightarrow (CH_3)_2COH^+ + PH_3$	189	364)
3	$CH_3CHOH^+ + PH_3 \rightarrow PH_4^+ + CH_3CHO$	182	364)
4	$H_3O^+ + PH_3 \rightarrow PH_4^+ + H_2O$	164	360,362)
5	$C_2H_5^+ + PH_3 \rightarrow PH_4^+ + C_2H_4$	158	360)
6	$CH_5^+ + PH_3 \rightarrow PH_4^+ + CH_4$	124	363)

The reactions 2 and 3 in Table 5 set the limits for the proton affinity of phosphine. Reaction 2 shows that it is smaller than that of acetone (186 ± 3 kcal/mol) and reaction 3 indicates that it is larger than the proton affinity of acetaldehyde (185 ± 3 kcal/mol). Thus a value of 185 ± 4 kcal/mol is obtained for the proton affinity of phosphine at room temperature.

Using the same methods [359,360] Eyler [354] obtained concurring results. Also the value reported by Haney and Franklin [361] of 186 ± 1 kcal/mol is in agreement. The latter workers determined the proton affinity of ammonia as 207 kcal/mol [362], which is approximately 21 kcal/mol larger than that for phosphine. The greater basicity of ammonia as compared to phosphine is shown by the difference of about 20 pH units in their relative basicities in aqueous solutions. The difference in the basicities of the aqueous solutions of 23 – 32 kcal/mol, which is comparable to that in the gas phase, leads to the somewhat surprising conclusion that solvent effects appear to play an unessential part in the relative basicities of PH_3 and NH_3 in aqueous solutions. The proton affinities of H_2O and H_2S, 164 and 170 kcal/mol, respectively, are in the reverse order.

The bond strengths can be obtained from the proton affinities. The proton and hydrogen affinities of a molecule and its respective ion are related to the ionisation potentials according to Eq. (12)

$$PA(M) - HA(M^+) = IP(H) - IP(M) \qquad (12)$$

The hydrogen affinity $HA(M^+)$ is simply the $H-M^+$ bond strength $D(H-M^+)$. A summary of the hydrogen affinities for phosphine and some isoelectronic molecules is given in Table 6.

Table 6. Bond strengths in PH_3, PH_4^+ and isoelectronic molecules and ions[a]

Ion	Proton affinity[b,365] [kcal/mol]	Molecule	Bond strength [366] [kcal/mol]
NH_4^+	128	CH_4	103
PH_4^+	102	SiH_4	80
OH_3^+	141	NH_3	105
SH_3^+	97	PH_3	85

[a] For literature for the values in the Table, see [359].
[b] From Eq. (12).

The substitution of a hydrogen atom in phosphine by an organic group increases the basic properties so that, for example, trimethylphosphonium salts are stable in aqueous solutions [78].

2. Toxicity of Phosphine

Phosphine is extremely poisonous. The maximum concentration of phosphine in the atmosphere must not exceed 0.1 ppm for an 8 h working day [456]. The smell of phosphine is first noticeable when the concentration reaches or exceeds about 2 ppm. A concentration of 50 – 100 ppm can only be withstood for a very short time without damage, a concentration of 400 ppm leads rapidly to death [225-227]. The symptoms observed by a medium-to-serious case of phosphine poisoning are: sense of anxiety, feeling of pressure in the chest, shortage of breath, pain behind the breast-bone, occasional dry cough, increased breathing noise, confusion, vertigo, and fainting. As first aid, the victim should be removed to fresh air and, when possible, given oxygen. For the toxicology of phosphine, see [225,226,228,229], as well as the article with an extensive bibliography by O. R. Klimmer [230]: Zur Frage der sog. chronischen Phosphorwasserstoffvergiftung.

3. Determination of Phosphine

Small quantities of phosphine in the air or other gases can be detected by passing the gas into a 5% aqueous solution of $HgCl_2$. The HCl, liberated by the formation of $P(HgCl)_3$, is then titrated potentiometrically [329]. The titration can be followed automatically using a Beckman Model K titrator [367]. A method for the semiquantitative determination of phosphine in the air works on the same principle [330]. Here the air is lead through 5 ml of a 1.5% aqueous $HgCl_2$ solution at pH 4.2. The solution is mixed with an indicator and the volume necessary to reduce the pH from 4.2 to 3.4 measured. By using a calibration curve, phosphine contents between 0.1 and 2.5 mg PH_3/litre of air can be estimated within an accuracy of ± 5%. A larger concentration range of 0.03 to 150 ppm PH_3 can be measured using a simple method. A fixed amount of air is sucked through a tube containing silica gel impregnated with $AuCl_3$. The phosphine concentration can be estimated from the length of the coloured zone [331].

Moser and Brukl [98] described a method for the gravimetric determination of phosphine. Dumas [368] used gas chromatography for the micro-determination of phosphine in the air (0.005–0.5 mg/litre), see also [369]. An automatic gas analyser, for the determination of phosphine and other substances in gases, works on the principle of the light absorption in reflection through a paper band, on which the gas causes a colour reaction with suitable reagents [370].

III. Preparation of Phosphine

For the laboratory preparation of phosphine, only a few of the many methods of formation are suitable. Among these the hydrolyses of calcium phosphide [79,80-83], magnesium phosphide [84-87], aluminium phosphide [88], zinc phos-

17

phide and tin phosphide [84-86,89,90] are the most important. As well as water, acids or bases, aqueous mixtures of acids or bases with organic solvents such as, for example, dioxane, alcohols etc. can be used for the hydrolyses [91-93].

Together with phosphine, noticeable amounts of diphosphine and higher phosphines are formed by the hydrolysis of calcium phosphide; thus, this reaction can be used for the preparation of such compounds [94]. Quesnel [89] reported that the formation of diphosphine can be avoided when aqueous hydrochloric acid is added drop-wise to a mixture of calcium phosphide and copper chloride (proportions by weight, $Ca_3P_2 : CuCl_2 = 10 : 1$) in boiling alcohol, for example, methanol, or in dioxane.

When the calcium phosphide is formed by the reduction of $Ca_3(PO_4)_2$ with carbon, the phosphine obtained on hydrolysis usually contains up to 3% acetylene.

Baudler and her co-workers [440] have described in detail the preparation of larger quantities of phosphine by the hydrolysis of calcium phosphide. Higher phosphines (see page 51) formed simultaneously, are thermally decomposed to phosphorus, phosphine and hydrogen. It is noteworthy that, on storage in steel cylinders the diphosphine concentration in phosphine, originally less than 1%, increases. It is, even after several months, so small that the gas is not spontaneously inflammable in contact with air, whereas, after about one year, it is spontaneously inflammable.

Our experience [95] has shown that the hydrolysis of aluminium phosphide with cold water is the most suitable method for the laboratory preparation of phosphine. Here it is important that the aluminium phosphide be as pure as possible in order to avoid the formation of spontaneously inflammable phosphine. The presence of small quantities of diphosphine and also higher phosphines are responsible for this spontaneous inflammability [96,276-278]. It appears, however, that these are only formed when P—P bonds are already present in the phosphide. Accordingly the hydrolysis of aluminium phosphide, prepared from the elements with phosphorus in slight excess, always leads to spontaneously inflammable phosphine. The formation of diphosphine and higher phosphines from aluminium or alkaline earth metal phosphides, which contain excess phosphorus, can be easily understood when the lattices of these compounds are considered.

Aluminium phosphide crystallises in a zinc blende lattice [441-443]. Both the aluminium and phosphorus atoms have a coordination number of four. The simplest picture of this is to consider the phosphide ions as forming a face-centred cubic lattice that is almost closest-packed with the aluminium ions occupying alternate tetrahedral sites. Only half of the tetrahedral sites are occupied as the compound has 1 : 1 stoichiometry and there are two tetrahedral sites associated with each phosphide ion. Because the aluminium ions occupy tetrahedral sites, it is clear that they have the coordination number four.

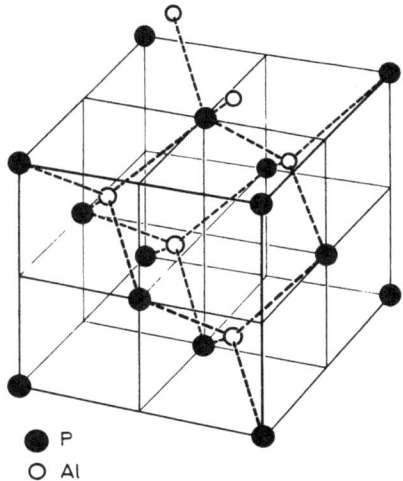

Fig. 3. Lattice of AlP (zinc blende lattice)

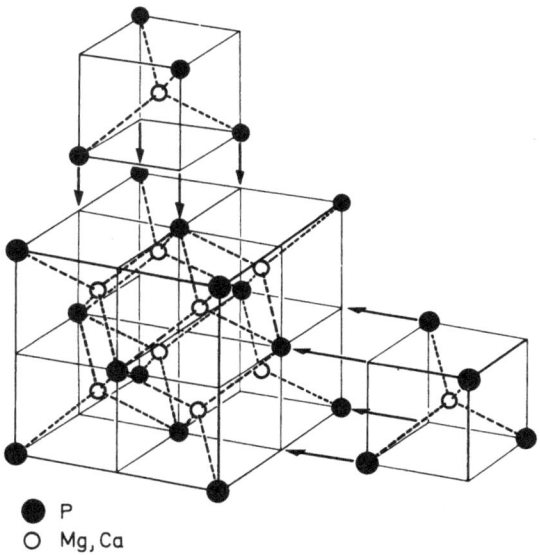

Fig. 4 Lattice of Ca_3P_2 and Mg_3P_2 (anti fluorite lattice with vacant sites)

Calcium phosphide and magnesium phosphide, crystallise in a lattice which can be deduced from the fluorite structure [444-446]. One way to describe the arrangement of the atoms in this lattice is to imagine that the phosphide ions form a face-centred cubic lattice and the calcium or magnesium ions occupy all tetrahedral sites. This structure is related to that of aluminium phosphide in which only half the tetrahedral sites in a face-centred cubic lattice are occupied. The lattice so far described yields, however, a stoichiometry of 2 : 1 or 4 : 2. Mg_3P_2 and Ca_3P_2 have, in fact, a stoichiometry of 3 : 2. This stoichiometry is obtained by vacant sites in the partial lattice of the metal ions. One quarter of the metal ion sites in the lattice are unoccupied.

Preparation of Phosphine by the Hydrolysis of Aluminium Phosphide

The generator consists of a 1-litre-3-necked-flask, fitted with a gas-inlet tube, and which is mounted over a magnetic stirrer. A second neck of the flask is connected to a trap, cooled in dry ice/methanol, *via* an approximately 2 metre long drying tube filled with P_4O_{10}/glass wool. A wash-bottle containing concentrated H_3PO_4 is fitted between the generator and the drying tube to serve as a control of the gas flow. It is important that the inlet-tube of the wash bottle projects only a little way below the surface of the phosphoric acid.

The generator flask is filled with 650 ml of water. The whole apparatus is then purged with nitrogen. Finally, with vigorous stirring and the introduction of moderate flow of nitrogen, 20 g of finely powdered aluminium phosphide are added. After a few minutes a continuous stream of phosphine is generated. The end of the PH_3-evolution is recognised by the change in colour of the suspension from green to grey. When necessary a further 20 g portion of aluminium phosphide can be added.

Using the above given conditions, an explosive decomposition of phosphine has never been observed. It is important, however, that no rapid change of pressure, which causes a spontaneous decomposition of phosphine into the elements, occurs because of the method of taking up of the phosphine or because of a blockage in the drying tube.

The hydrolyses of aluminium phosphide with acids [89,97,98] or alkalies [99, 100] do not seem to be so suitable for the preparation of phosphine.

The method, described relatively early, for the preparation of phosphine using the reaction of hot concentrated alkalies, such as NaOH, KOH or $Ca(OH)_2$, on white phosphorus can also be used for the laboratory preparation. This method also produces a steady stream of phosphine, which, however, may be contaminated by up to 90% hydrogen and traces of P_2H_4 [1,3,85,101-106]. Phosphine formed from the thermal decomposition of phosphorous or hypophosphorous acids or their salts is similarly contaminated with hydrogen. In cases where hydrogen interfers, the phosphine can be purified by condensation and distillation.

"Inorganic Syntheses", Vol. IX, p. 56, contains an accurate description of the preparation of phosphine by the pyrolysis of phosphorous acid [107,108].

Very pure phosphine is formed by the hydrolysis of phosphonium iodide with water, dilute acids or dilute bases [2,14,17,107,109-114], or by the reduction of phosphorus trichloride with lithium in diethyl ether [87,115-117]. Related to the latter is a method for the preparation of phosphine, described in the patent literature, where phosphorus trichloride vapour, diluted with nitrogen, is passed through a column filled with lithium hydride mixed with an inert material, such as sand, NaCl, KCl or similar materials [118].

According to a method described by Horner et al., phosphine was obtained in 70% yield from the reaction of PCl_3 with finely divided sodium in toluene, followed by hydrolysis of the reaction products [287].

Finally, in the recent patent literature, some further processes for the preparation of phosphine were described; for example, the treatment of white phosphorus with steam in the presence of phosphoric acid at 275–285 °C. According to a British patent, phosphine is formed when white phosphorus, in aqueous acid, is brought into contact with mercury or zinc amalgam [119]. A Japanese patent recommends the treatment of a mixture of white phosphorus and granulated zinc with acids and a small amount of methanol for the preparation of highly pure phosphine [371]. Other patents describe electrolytic processes. Finally, it is mentioned that phosphine is formed by the electrolysis of phosporous and hypophosphorous acid, especially at mercury or lead cathodes [120].

The purification of phosphine, from the main gaseous substances obtained by the preparation, can be achieved by fractional distillation [14,121,122]. Acetylene can be removed with the help of molecular sieves [123].

IV. Reactions of Phosphine

1. Thermal Decomposition, Reaction with Oxygen, Reducing Properties

Phosphine decomposes only at higher temperatures. The thermal decomposition of phosphine under various conditions has been examined in detail. It is found to be a first order reaction. The rate constant for the decomposition of phosphine at 500 °C is approximately $8 \cdot 10^{-3}$ sec^{-1} [57,231-233].

The thermal decomposition of phosphine on a glass surface can be satisfactorily described as a first order reaction using Eq. (13):

$$-\frac{dp}{dt} = K \cdot p \qquad (13)$$

where K is the rate constant for the decomposition and p is the pressure of phosphine. The decomposition rate on a silicon film is smaller than that on a glass surface. This decomposition also follows first order kinetics. The temperature dependence is described by the Arrhenius equation:

21

$$\lg K = \ 8.86 - \ 9669 \cdot \frac{1}{T} \qquad \text{(glass surface)} \qquad (14)$$

$$\lg K = 11.6 \ - 12110 \cdot \frac{1}{T} \qquad \text{(silicon surface)} \qquad (15)$$

The energies of activation of the decomposition on glass and silicon surfaces are 44.2 and 55.3 kcal/mol, respectively [372]. See also [373-375, 481-492].

Phosphine-oxygen mixtures are relatively stable above the upper and below the lower critical explosion pressures. However, on irradiation with UV-light reaction occurs. It is assumed that the reaction is initiated by the photolytic dissociation of the PH_3 molecule according to the equation:

$$PH_3 + h \cdot \nu \ \rightarrow \ PH_2 + H \qquad (6)$$

Then the PH_2 radical reacts further by collision with an oxygen molecule. In the presence of water vapour the end-product is H_3PO_2 and/or H_3PO_4 [477-480].

The conditions under which mixtures of phosphine and oxygen ignite were investigated by Trautz [234] and Shantarovich [235]. The ignition pressure is dependent on the composition of the mixture, the water content, the temperature and the presence of foreign gases. With increasing partial pressure of phosphine the ignition pressure generally increases. Mixtures of phosphine and excess oxygen diluted with nitrogen do not react noticeably up to temperatures of 200 °C [376].

A large number of articles have been devoted to kinetic studies of the oxidation of phosphine [236-243]. The compositions of phosphine under pressure, which are oxidised by water, have been described by Bushmakin and Frost [244].

Mixtures of phosphine and oxygen, both above and below the explosion limits, subjected to flash photolysis show, in the spectra, the presence of PH-, OH- and PO-radicals as well as the PH_2-radical [255]. During the reaction of atomic oxygen with phosphine visible luminescence up to 3600 Å and UV emission were observed, which were attributed to the partial processes: $O + PO \rightarrow PO_2$ and $OH + PO \rightarrow HOPO$ [377].

It is of particular interest that solutions of hydrogen peroxide of varying concentrations are not able to oxidise phosphine [256]. Phosphine appears to simply dissolve (to a very small extent) even in 100% H_2O_2 without reacting [282]. Solid sulphur trioxide oxidises phosphine to red phosphorus [283]. No reactions are observed with NO and N_2O under the usual conditions [79,285,286].

The reducing action of phosphine has been used in organic chemistry in various ways. The reactions described in the literature are summarised by the following equations:

$$p-CH_3 \cdot C_6H_4SO_2NNaCl \ \longrightarrow \ p-CH_3 \cdot C_6H_4SO_2NH_2 \ \ [245] \qquad (16)$$

$$\alpha\text{-naphthol} \ \longrightarrow \ \text{naphthalene} \ [246] \qquad (17)$$

$$C_6H_5NO_2 \longrightarrow C_6H_5NH_2 \text{ [247)]} \tag{18}$$

A comprehensive study of the reducing properties of phosphine particularly with respect to aromatic nitro-compounds and aromatic sulphonyl chlorides, was published by Buckler et al. [248)].

2. The Question of the Existence of Phosphine Oxide OPH₃

In contrast to organo-phosphines and also many other phosphorus (III) compounds, phosphine is very unreactive and shows no tendency to react with electrophilic substances. In particular there is no substantiated evidence for an auto-oxidation of phosphine. This chemical behaviour is related to the electronic configuration of the molecule. As shown by, among other factors, H–P–H bond angles and the chemical shift, δ_{31p}, in the nuclear magnetic resonance spectrum, the lone electron pair of the phosphine molecule stays predominantly in a $3s$ orbital of the phosphorus atom. The relative proximity of these two electrons to the nucleus and their, to a first approximation, spherosymmetric density distribution cause the unusually strong shielding of the phosphorus nucleus and thus the high positive chemical shift of $+241$ ppm (relative to 85% orthophosphoric acid) and also the small nucleophilic character of the molecule. Derivatives of phosphine which also show high positive chemical shifts in the ^{31}P NMR spectra are collected together in Table 7. In common with phosphine, they do not react either with oxygen or with sulphur to form the corresponding oxides or sulphides, nor do they react with alkyl iodides to form the corresponding phosphonium salts. The resonance signal is shifted to lower field strengths only when one or more ligands in phosphine, tris-(trimethylsilyl)-, tris-(trimethylstannyl)-, tris-(trimethylgermanyl) - or tris-(triphenylstannyl)-phosphine are substituted. This is shown in the table, for example, by the series of compounds $P[Si(C_6H_5)_3]_3$, $P[Si(C_6H_5)_3]_2(C_6H_5)$, $P[Si(C_6H_5)_3](C_6H_5)_2$ and $P(C_6H_5)_3$. Parallel to this, the nucleophilic character increases, thus the usual reactivity of phosphorus(III) compounds towards electrophiles is reached step-wise. The contribution of the s electrons to the s bonding system is obviously increasing in the series, so that the orbital occupied by the electron lone pair has more and more p character. This orbital is more far-reaching and thus gives the molecule increasing nucleophilic character [393)]. In agreement with this argument is the increased reactivity of phosphine in UV-light caused by the promotion of one electron of the lone pair to a more outlaying orbital. This will be discussed in more detail later.

In the course of mass spectroscopic investigations of the hydrolysis products of calcium phosphide, Baudler and her co-workers could find no evidence for the existence of OPH₃ as an oxidation product of PH_3 [396)]. In contrast, oxides of higher phosphines were observed, even when the hydrolysis of calcium phosphide was carried out with the strictest exclusion of oxygen. The

Table 7. N.M.R. data for phosphorus, phosphine and derivatives of phosphine

Compound	Chemical shift δ_{31P} [ppm]	Coupling constant J_{HP} [Hz]	Lit.
P_4, solid	460.0		378,383)
P_4, solution	460.0–533		378–382)
P_4, vapour	553.1		384)
PH_3	238	188.2	385–391)
$P[Si(CH_3)_3]_3$	251.2		392)
$P[Sn(CH_3)_3]_3$	330.0		392,395)
$P[Ge(CH_3)_3]_3$	228.0		392,395)
$P(SiH_3)_3$	378.0		339)
$PH[Si(CH_3)_3]_2$	237.4	186	394)
$PH_2[Si(CH_3)_3]$	239.0	180	394)
$P[Sn(C_6H_5)_3]_3$	323		395)
$P[Sn(C_6H_5)_3]_2(C_6H_5)$	163		395)
$P[Sn(C_6H_5)_3](C_6H_5)_2$	56		395)
$PH_2(CH_3)$	163.5		383)
$PH(CH_3)_2$	98.5		383)
$P(CH_3)_3$	62.0		383)
$PH_2(C_6H_5)$	122.0		383)
$PH(C_6H_5)_2$	41.1		383)
$P(C_6H_5)_3$	8.0		383)
$P[C(CH_3)_3]_3$	– 58		447)
$P[Ge(CH_3)_3]_2CH_3$	177		447)
$P[Ge(CH_3)_3]_2C_6H_5$	127		447)
$PH[Ge(CH_3)_3](C_6H_5)_2$	119		447)
$P(GeH_3)_3$	338		448)
$P[Sn(C_4H_9)_3]_2C_6H_5$	170		448)
$P[Sn(C_4H_9)_3](C_6H_5)_2$	56		448)

following phosphine oxides were identified individually: P_2H_4O, P_3H_3O, P_3H_5O, P_4H_2O, P_7H_3O and P_7H_5O. PH_3O was not observed. The oxygen of the higher phosphine oxides was probably contained in the calcium phosphide used for the hydrolysis [396].

Also, attempts to convert phosphine to PH_3O using amine oxides, such as trimethylamine oxide and pyridine oxide, did not proceed to the result formulated in Eq. (19) [397]:

$$PH_3 + ONR_3 \xrightarrow{\;\;|\!\!\!\!\!\nrightarrow\;\;} OPH_3 + NR_3 \tag{19}$$

Phosphine reacts neither in aqueous solution nor direct with trimethylamine oxide. Also no reaction occurs on treatment of phosphine with pyridine oxide in methylene chloride [397]. In contrast, PCl_3 reacts rapidly and quantitatively with pyridine oxide in methylene chloride even at $-40\ °C$ to give $OPCl_3$ and pyridine [398].

The reaction products observed from experiments to reduce the phosphoryl halides, $OPCl_3$ and $OPBr_3$, with lithium hydride suggest that the primary reactions proceed according to Eqs. (20) and (21):

$$OPCl_3 + 3\ LiH \longrightarrow OPH_3 + 3\ LiCl \qquad (20)$$

$$OPBr_3 + 3\ LiH \longrightarrow OPH_3 + 3\ LiBr \qquad (21)$$

The authors assume that OPH_3 rearranges to the tautomeric form H_2POH, from which, at temperatures below $-115\ °C$, the highly polymeric $(PH)_x$ is formed by condensation:

$$\text{HO} - \overset{\displaystyle |}{\underset{\displaystyle H}{P}} - H + \text{HO} - \overset{\displaystyle |}{\underset{\displaystyle H}{P}} - H + \text{HO} - \overset{\displaystyle |}{\underset{\displaystyle H}{P}} - H \longrightarrow \overset{\displaystyle |}{\underset{\displaystyle H}{P}} - \overset{\displaystyle |}{\underset{\displaystyle H}{P}} - \overset{\displaystyle |}{\underset{\displaystyle H}{P}} - + 3\ H_2O$$

$$(22)$$

The released water reacts with excess lithium hydride to give hydrogen.

$$LiH + HOH \longrightarrow LiOH + H_2 \qquad (23)$$

Thus, in the case of the reaction with $OPBr_3$, the reaction shown in Eq. (24) was observed.

$$OPBr_3 + 4\ LiH \longrightarrow PH + H_2 + 3\ LiBr + LiOH \qquad (24)$$

When lithium aluminium hydride, $Li(AlH_4)$, in ether solution is used instead of lithium hydride for the reduction of $OPCl_3$, the OPH_3 also formed is mostly further reduced to phosphine at temperatures of about $-115\ °C$. Altogether, the reaction can be described by the following equations:

$$4\ OPCl_3 + 5\ Li(AlH_4) \begin{array}{c} \xrightarrow{\ >90\%\ } 4\ PH_3 + 3\ Li(AlCl_4) + 2\ LiAlO_2 + 4\ H_2 \\ \xrightarrow{\ <10\%\ } 4\ PH + 3\ Li(AlCl_4) + 2\ LiAlO_2 + 8\ H_2 \end{array} \qquad (25,26)$$

In the reaction of $OPCl_3$ with lithium borohydride at $-115\ °C$ proceeding according to Eq. (27)

$$OPCl_3 + 3\ Li[BH_4] \longrightarrow H_3PO \cdot BH_3 + 3\ LiCl + 2\ BH_3 \qquad (27)$$

a crystalline solid $H_3PO \cdot BH_3$ is formed, which is stable up to $-90\ °C$. Thus, OPH_3 in the form of the adduct $H_3PO \cdot BH_3$ can be isolated and is stable up to $-90\ °C$, while, in the free state, it decomposes even at $-115\ °C$ according to $PH_2OH \rightarrow PH + HOH$. Above $-90\ °C$, $H_3PO \cdot BH_3$ also decomposes to hydrogen and H_2POBH_2, which is regarded by the authors as polyborophosphen oxide $-\overset{+}{P}H_2-O-\overset{-}{B}H_2-O-\overset{+}{P}H_2-O-\overset{-}{B}H_2-$ [397].

SCF calculations for the hypothetical compound H_3PO were carried out by Marsmann et al. [449]. The authors studied the effect of adding either a d or another p orbital to a phosphorus atom depicted in terms of seven s and three p Gaussian orbitals. An approximate model of H_3PO based on the valence bond method has been published by Mitchell [450].

3. Reactions with Atomic Hydrogen and Nitrogen

The reaction between phosphine and deuterophosphine on one hand and atomic hydrogen, generated by high-voltage discharge, on the other hand was intensively investigated using spectroscopic methods by Guenebaut and Pascat [249-252]. According to Wiles and Winkler [253], the stable end products of the reaction are red phosphorus and molecular hydrogen.

The main products from the reaction between phosphine and atomic nitrogen are molecular hydrogen and phosphorus nitride $(PN)_x$, which is formed in the α form [253]. For the conversion of one mole of phosphine into (PN), two atoms of nitrogen appear to be necessary. The primary step is the formation of PH_2 radicals which react further with the nitrogen atoms to give the nitride [254].

4. The Systems Phosphine/Water, Phosphine/Water/Ammonia, and Phosphine/Water/Ammonia/Methane

Electric discharges in the system phosphine/water result in the formation of water insoluble polyphosphines $H_2P-(PH)_n-PH_2$ and phosphorus as well as H_3PO_2, H_3PO_3 and H_3PO_4. When ammonia is also present, hypophosphate, pyrophosphate, polyphosphate and possibly polyhypophosphate are formed in addition. Finally, in the system phosphine/water/ammonia/methane, organic phosphorus compounds such as aminoalkylphosphates and aminoalkanephosphonates and other phosphorus-free compounds such as amino-acids, ethanolamine etc. can be detected as well as the previously mentioned reaction products. The presence of phosphine or its rearrangement products seem to make the condensation reactions in the last mentioned system possible, because the ratio of the amino-acids, which are present after hydrolysis with 6N HCl, to the amino-acids which are present before the hydrolysis is larger in this system than in the system $CH_4/H_2O/NH_3$ [399,400].

Irradiation of a gaseous mixture of phosphine and ammonia with ^{60}Co gamma-rays produces hydrogen, nitrogen and red phosphorus. Other products were not observed. Phosphine appears to work as a very efficient radical acceptor via reactions such as

$$NH_2\cdot + PH_3 \longrightarrow NH_3 + PH_2 \tag{28}$$

and

$$H\cdot + PH_3 \longrightarrow H_2 + PH_2\cdot \tag{29}$$

The phosphino radical reacts further and finally produces red phosphorus and hydrogen [401,402].

5. Reactions with Alkali and Alkaline Earth Metals

Alkali metals react with phosphine to form alkali dihydrogen phosphides. The reactions are generally carried out by passing the phosphine through a solution of the alkali metal in liquid ammonia [124-129]. According to a patented method, phosphine can also be led through a suspension of the alkali metal in a mixture of a hydrocarbon and methoxypolyether [130]. Finally, sodium dihydrogenphosphide is also formed on leading phosphine into an ether solution of triphenylmethyl sodium [131].

Step-wise replacement of all three hydrogen atoms in phosphine by lithium can be obtained by the reaction of phosphine with an alkyl lithium compound in the corresponding molar ratio. The preparations of Li_3P and Na_3P from the elements were described by Brauer and Zintl [134].

Of the alkali dihydrogenphosphides, potassium dihydrogenphosphide, KPH_2, has been studied most extensively. This compound crystallises in a rock salt lattice with pseudo-rhombic distortion in the direction of a space diagonal [132]. An analogous structure has also been proved for rubidium dihydrogenphosphide. The density of potassium dihydrogenphosphide at 25 °C is 2.435 g/cm^3 [132]. Potassium dihydrogenphosphide is soluble in strongly polar solvents such as tetramethylene sulphone and dimethylacetamide [133]. It is also well soluble in trimethylamine and dimethylformamide (10–15 g/100 ml). The chemical shift, δ_P, of KPH_2 in liquid ammonia is 279 ± 2 ppm (relative to 85% aqueous orthophosphoric acid), the coupling constant is J_{PH} = 139 Hz [95]. The proton nuclear resonance spectrum of KPH_2 dissolved in dimethylformamide consists of a doublet with a chemical shift δ_H = 1.12 ppm (relative to $(CH_3)_4Si$), the phosphorus spectrum consists of a triplet with a chemical shift of + 255.3 ppm (relative to 85% aqueous orthophosphoric acid). The coupling constant, measured from the proton spectrum, is J_{HP} = 136.8 Hz [133]. Solutions of KPH_2 in trimethylamine show a well resolved triplet in the phosphorus spectrum with a chemical shift of 272 ± 2 ppm [95]. The ^1H and ^{31}P resonance spectra of solutions of $NaPH_2$ and KPH_2 in liquid ammonia were thoroughly investigated by Sheldrick [144]. As previously mentioned, the proton spectra consist of doublets and the phosphorus spectra of 1 : 2 : 1 triplets. When considerable quantities of phosphine are present in the solutions the multiplet structures collapse. The chemical shift of a 4% solution of $NaPH_2$ in liquid ammonia at room temperature was measured as 11.487 ± 0.004 τ. This depends only slightly on temperature and is, at 2 °C 11.469 ± 0.003 τ, and at – 23 °C is 11.444 ± 0.004 τ. The coupling constant, J_{HP}, at + 22 °C is 138.71 ± 0.007 Hz. This increases with decreasing temperature and at 2 °C is 138.86 ± 0.05 and at – 23 °C it is 139.05 ± 0.07 Hz. Measurements of a 2% solution of $NaPH_2$ in liquid ammonia at $-60\,°C$ gave a value of 140.08 ± 0.04 Hz. The ^{31}P spectrum of a 5% solution of KPH_2 in liquid ammonia at $22\,°C$ shows a chemical shift of $+393$ ppm (relative to an external P_4O_6 reference). The coupling constant obtained from this spectrum is $J_{PH} = 139 \pm 2$ Hz.

Potassium dihydrogenphosphide dissolved in dimethylformamide, is probably dissociated into ions. This is confirmed by the equivalent conductance, the value of $77.0\ \Omega^{-1}\,cm^2\,mol^{-1}$ was found for a $8 \cdot 10^{-5}\ mol/cm^3$ solution. The PH_2^- ion is a strong nucleophilic reagent. The study of its chemical behaviour towards oxygen, sulphur and white phosphorus has produced, to date, no conclusive results. The reaction with the latter element results in the formation of an amorphous red-brown substance of composition KP_5H_2 which is soluble in dimethylformamide [133].

Li_3P and Na_3P have the same structure as Na_3As. Each phosphorus atom is surrounded by 5 alkali metal ions at the corners of a trigonal bipyramid. The lattice contains two types of alkali metal atoms. One sort is surrounded by a trigonal prism of other alkali metal ions, in which the centres of the three vertical faces of the prism are occupied by three phosphorus atoms. The other type of alkali metal ion is surrounded by 4 phosphorus atoms in a distorted tetrahedron. Seven alkali metal ions are found at larger distances. In sodium phosphide the distance between the central phosphorus atom of the trigonal bipyramid and an axial sodium atom is 2.93 Å and that between the phosphorus atom and an equatorial sodium atom 2.88 Å [134].

As well as for the preparation of alkali phosphides, the reaction of phosphine with the elements, their oxides or halides, at higher temperatures in quartz tubes have been much used recently for the preparation of other phosphides, in particular those which play important roles in semi-conductor technology. The preparations of the following phosphides using these methods have been described: for example, NdP [135,136], BP [137,138], GaP [139,140], SmP, LaP [136,141], TiP, Ti_2P (possibly Ti_3P) [142,143] and InP [139]. See also Section IV.9.

Calcium reacts with phosphine in an analogous manner as the alkali metals. In liquid ammonia, solid $Ca(PH_2)_2 \cdot nNH_3$ is formed with hydrogen evolution [128,280]. The corresponding reaction with a solution of elemental strontium in liquid ammonia does not lead to a uniform product [280].

6. Reactions with Halogens and Chloramines

Phosphine burns to PCl_5 and hydrogen chloride in a chlorine atmosphere [79,311,313]. An aqueous chlorine solution oxidises PH_3 to phosphoric acid (see also Section IV.7) [314].

Royen and Hill [303] reported that phosphonium bromide and solid phosphorus hydride are the products from the reaction of excess phosphine with bromine at low temperatures.

From iodine and dry phosphine, P_2I_4 and hydrogen iodide, which reacts with excess phosphine to form phosphonium iodide, are formed [316,317].

In aqueous solution iodine reacts with phosphine according to the empirical formula (30) to form phosphorous acid and hydrogen iodide [220];

$$PH_3 + 3\,I_2 + 3\,H_2O \longrightarrow H_3PO_3 + 6\,HI \qquad (30)$$

According to earlier work [221], hypophosphorous acid should be formed according to Eq. (31):

$$PH_3 + 2\,I_2 + 2\,H_2O \longrightarrow H_3PO_2 + 4\,HI \qquad (31)$$

In fact, this is the first stage of the reaction. In the presence of protons released by the reaction, oxidation to phosphorous acid occurs. If the reaction mixture is kept only weakly acid throughout the whole reaction, only hypophosphite is actually formed in the solution.

Horak and Ettel [222] described two methods for the determination of phosphine in organic solvents which are based on Eq. (31). Thus,

        a) PH_3 is oxidised with excess iodine in a bicarbonate containing solution to H_3PO_2 and the excess iodine is back titrated with sodium thiosulphate solution;

        b) PH_3 mixed with twelve volumes of methanol is titrated with an aqueous iodine solution in the presence of pyridine so that H_3PO_2 is formed.

Whereas the reactions of chloramines of the type R_2NCl (where R = H or an alkyl group) with tertiary phosphines have been abundantly investigated and almost without exception lead to the formation of aminophosphonium chlorides or their condensation products [403-407], Highsmith and Sisler [408] investigated the behaviour of primary and secondary phosphines and phosphine itself towards chloramines for the first time. With dimethylchloramine, phosphine forms red phosphorus and dimethylammonium chloride according to Eq. (32):

$$2\,PH_3 + 3\,(CH_3)_2NCl \longrightarrow 2P + 3\,[(CH_3)_2NH_2]Cl \qquad (32)$$

Only polymeric phosphorus hydride, P_xH_y, and not red phosphorus was observed when phosphine was treated with chloramine in ether solution [408,409]. The authors attributed this to the expected higher basicity of a dimethylamino group in comparison to that of an amino group.

In an alkaline medium phosphine is oxidised solely to hypophosphite by N-bromosuccinimide [410].

7. Reactions with Sodium Hypochlorite

Sodium hypochlorite in aqueous solutions reacts practically instantaneously with phosphine so that such solutions are particularly suitable for removing traces of phosphine from a gas stream [219].

In the pH range 12–13 the reaction follows the equation

$$PH_3 + 2\,NaOCl \longrightarrow H_3PO_2 + 2\,NaCl \tag{33}$$

The reaction is first order with regard to both the concentration of phosphine and the concentration of hypochlorite. It is suggested that the reaction proceeds as shown in Eqs. (34) – (36):

$$OCl^- + H_3O^+ \rightleftharpoons HOCl + H_2O \qquad \text{fast} \tag{34}$$

$$PH_3 + HOCl \longrightarrow (PH_3O) + Cl^- + H^+ \qquad \text{rate-determining} \tag{35}$$

$$(PH_3O) + OCl^- \longrightarrow H_3PO_2 + Cl^- \qquad \text{very fast} \tag{36}$$

8. Reactions with Hydrogen Halides

The introduction of phosphine into liquid hydrogen chloride results in the formation of PH_4Cl [295,328]; this was also observed earlier [307,308]. Similarly PH_4Br can be obtained from hydrogen bromide and phosphine [307,309,328]. Both phosphonium halides are almost completely dissociated at room temperature and normal pressure.

The reaction between dry phosphine and hydrogen iodide, first described in 1817 by J. J. Houton de la Billardière [109] produces phosphonium iodide. The simplest laboratory preparation of this compound is by the hydrolysis of an intimate mixture of diphosphorus tetraiodide and white phosphorus [279]. According to X-ray diffraction investigations, phosphonium iodide crystallises in a caesium chloride type lattice [324,325]; see also [326]. The hydrogen atoms are tetrahedrally arranged about the phosphorus atoms; this was also shown by Raman and IR spectra [327,328].

9. Reactions with Metal and Non-Metal Halides

With metal halides such as $AlCl_3$ or InI_3 [93,146] and non-metal halides such as BCl_3 or BBr_3 phosphine forms 1 : 1 adducts which, on pyrolysis, split off hydrogen halides to form the corresponding phosphides [147]. This method for the preparation of phosphides is used in practice for the formation of semiconducting crystals. The 1 : 1 adduct, $AlCl_3 \cdot PH_3$ reacts exothermically with alkyl halides to form hydrogen chloride and the primary phosphine. This reaction offers a relatively convenient method for the preparation of primary phosphines. The yields, especially with alkyl halides with long chain alkyl groups (more than 5 carbon atoms), are favourable and usually greater than 50% [93].

Phosphine reacts with B_2Cl_4 at room temperature according to equation (37)

$$B_2Cl_4 + 2\,PH_3 \longrightarrow B_2Cl_4 \cdot 2\,PH_3 \tag{37}$$

1

to form the colourless solid *1* which is stable up to 65 °C but is, however, very sensitive to air [334].

Tris-(triorganylmetal)phosphines are the products from the reactions between triphenyltin, triphenylgermanium and triphenyllead chlorides and phosphine [148,149]. The reactions proceed in benzene solution in the presence of triethylamine as hydrogen chloride acceptor thus:

$$3 (C_6H_5)_3MCl + PH_3 + 3 (C_2H_5)_3N \longrightarrow [(C_6H_5)_3M]_3P + 3[(C_2H_5)_3NH]Cl \qquad (38)$$

M = Sn, Ge, Pb.

To date, however, only few reactions between phosphine and a non-metal halide, in which a chemical bond is formed between phosphorus and a non-metal by HCl condensation, are known. To these, apart from the above-mentioned reactions, belongs also the reaction with CF_3SCl [223] which, depending upon the chosen proportions of the reactants, in a sealed tube at –95 °C leads to the formation of $(CF_3S)_2PH$ or $(CF_3S)_3P$ [224]. Both compounds are not very stable thermally and decompose at 40–50 °C. Tris(trifluoromethylthio)-phosphine forms an unstable adduct with chlorine, which decomposes at 0 °C to give a mixture of PCl_3, bis(trifluoromethyl)-disulphide and trifluoromethyl-sulphenyl chloride.

With SiH_3I phosphine probably forms silylphosphine [288] and with $(CH_3)_2BBr$, crystalline $H_2PB(CH_3)_2$ [289]. Germanium halides do not react with phosphine [299].

The phase diagram of the system PH_3/BF_3 indicates the presence of two compounds [290]. On treatment of PH_3 with BF_3 at –130 °C $BF_3 \cdot PH_3$ is formed [291,292], which is probably converted to $[PH_3(BF_2)]BF_4$ by the following reaction mechanism [290,292,293]:

$$PH_3 \cdot BF_3 \xrightarrow{-HF} PH_2(BF_2) \xrightarrow{+HF} PH_3(BF_2) F \xrightarrow{+BF_3} [PH_3(BF_2)]BF_4 \qquad (39)$$

In many other cases the course of the reaction and the nature of the reaction products resulting from the treatment of non-metal halides with phosphine are not completely clarified. To these belong the reaction products shown in Table 8.

For completeness, the alkylsilylphosphines, a class of compounds, the first member of which was synthesised by Parshall and Lindsey [150] must be mentioned. The alkylsilylphosphines, $(CH_3)_3SiPH_2$ (b.p. 69–73 °C), $[(CH_3)_3Si]_2PH$ (b.p. 170–172 °C) and $[(CH_3)_3Si]_3P$ (b.p. 242–243 °C), are formed by the reactions of the alkylchlorosilane with the lithium phosphides, $LiPH_2$, Li_2PH and Li_3P, e.g. [150].

$$Li_3P + 3 (CH_3)_3SiCl \longrightarrow 3 LiCl + [(CH_3)_3Si]_3P \qquad (40)$$

Leffler and Teach [332] described the preparation of $[(CH_3)_3Si]_3P$ from $NaPH_2$ and $(CH_3)_3SiCl$.

Table 8. Reaction products from phosphine and non-metal halides

Reaction of PH_3 with	Reaction conditions	Reaction product	Lit.
BCl_3	20 °C	$PH_3 \cdot BCl_3$	147,158,294)
B_2Cl_4	-78 °C	$2\,PH_3 \cdot B_2Cl_4$	296)
BBr_3	20 °C	$PH_3 \cdot BBr_3$	147,297)
BBr_3	1250 °C	BP	298)
SiF_4	-22 °C, 50 at		293)
$SiCl_4$	-60 °C	No reaction	299)
$SiBr_4$	Under pressure	Colourless amorphous compound	300)
PCl_3		Solid phosphorus hydride	301)
PCl_5	Dissolved in liq. HCl	No defined reaction	302,303)
PBr_5	-90 °C	PH_4Br, solid phosphorus hydride	301,303)
$AsF_3, AsCl_3$		Arsenous phosphide	304-306)

Bis-(trimethylsilyl)phosphine and trimethylsilylphosphine are also formed by the hydrolysis of tris-(trimethylsilyl) phosphine with the appropriate amount of water in diglyme or tetrahydrofuran [151]:

$$2\,[(CH_3)_3Si]_3P + HOH \longrightarrow 2\,[(CH_3)_3Si]_2PH + [(CH_3)_3Si]_2O \qquad (41)$$

$$[(CH_3)_3Si]_3P + HOH \longrightarrow (CH_3)_3SiPH_2 + [(CH_3)_3Si]_2O \qquad (42)$$

The compounds 2 and 3 were formed by treating a mixture of Li_2PH and Li_3P with diethyldichlorosilane [150]:

2   3

The alkylsilylphosphines are colourless liquids or low melting solids. They are very sensitive to water and oxygen but are thermally very stable.

Phosphine, primary and secondary phosphines react with Grignard reagents as "active hydrogen" compounds [214-216,493].

10. Reactions with Diborane and Monobromodiborane

Ammonia reacts with boron hydride to form a product with the empirical composition $B_2H_6 \cdot 2NH_3$ [152-154]. Similarly, phosphine combines with diborane, in the gas phase above $-30\,°C$ and in the liquid phase in a sealed tube above $-130\,°C$, to give a white instable solid of constitution BH_3PH_3 (m.p. 32-$35\,°C$) [155,156]. The dissociation pressure of $BH_3 \cdot PH_3$ is so large that the compound is already separated into its components at room temperature

$$BH_3 \cdot PH_3 \longrightarrow PH_3 + 1/2\, B_2H_6 \qquad (43)$$

Nuclear magnetic resonance spectra show that the compound exists as a monomer in the molten state; IR and Raman data show that the same molecular structure exists for the solid state [156]. Sawodny and Goubeau [411] calculated the force constants from the normal vibrations of the molecule, after they had corrected the original assignments of the bands [156]. A bond number of 0.78 was found for the P−B bond. The chemical shifts and coupling constants from the ^1H and ^{11}B n.m.r. spectra for molten BH_3PH_3 are given in Table 9 [260].

Table 9. N.M.R. data for molten BH_3PH_3 at $37\,°C$

Chemical shift [ppm]		Coupling constant [Hz]	
δ_{BH}	$+0.53$[a]	J_{BH}	104
δ_{PH}	$+4.31$	J_{PH}	372
δ_{B}	60.8[b]	J_{PB}	27
		J_{HBPH}	8
		J_{PBH}	16

[a] Relative to TMS.
[b] Relative to $B(OCH_3)_3$.

The ^1H n.m.r. and the vibrational spectra of $PH_3 \cdot BD_3$, $PD_3 \cdot BH_3$ and $PD_3 \cdot BD_3$ were investigated by Davis and Drake [412]. Proton n.m.r. spectra of the first two compounds show that no hydrogen-deuterium exchange takes place between boron and phosphorus. The spectrum of $BH_3 \cdot BD_3$ shows only a signal for the PH_3 group. This appears as a widely split doublet because of coupling between the protons and the phosphorus nucleus ($J_{PH} = 360$ Hz). Each component of the doublet is further split into a septet due to coupling between the protons bonded to phosphorus and the three deuterons ($I = 1$). The coupling constant J_{HD} is 1.1 Hz. The proton spectrum of $PD_3 \cdot BH_3$ consists only of a resonance for the protons on the boron atom ($I = 3/2$ for ^{11}B);

E. Fluck

this is comprised of a $1:1:1:1$ quartet (J_{BH} = 103 Hz). Each component is further split into a doublet due to coupling with the phosphorus nucleus.

Trimethylamine replaces PH_3 quantitatively in $BH_3 \cdot PH_3$ to give $BH_3 \cdot N(CH_3)_3$. Liquid or gaseous ammonia also replaces phosphine in the compound $BH_3 \cdot PH_3$ to the extent of 52–58% or up to 75%, respectively. The hydrogen atoms bonded to boron are partially, or completely, replaced by chlorine on treatment with hydrogen chloride, depending on the reaction conditions. The first stage of the reaction with hydrogen chloride is the formation of the colourless, viscous liquid $BH_2Cl \cdot PH_3$ (see below).

$$BH_3 \cdot PH_3 + HCl \longrightarrow BH_2Cl \cdot PH_3 + H_2 \qquad (44)$$

This reacts further with hydrogen chloride to give a colourless, crystalline product of composition $BHCl_2 \cdot PH_3$. The latter compound melts at about 68 °C. Finally, this is converted to $BCl_3 \cdot PH_3$ above 0 °C and under higher pressure. $BCl_3 \cdot PH_3$ is a colourless compound and was prepared earlier by Besson [157] from the reaction of equal volumes of PH_3 and BCl_3 below 20 °C; its properties were later investigated thoroughly by Tierney [158]. According to this work, $BCl_3 \cdot PH_3$ is a well-crystalline solid which is partly dissociated in the gas phase (at 31 °C to about 90%). The enthalpy of the reaction

$$Cl_3B_{(g)} + PH_{3(g)} \longrightarrow BCl_3 \cdot PH_{3\,(solid)} \qquad (45)$$

is 26 kcal/mol. In benzene the compound is probably monomeric and essentially undissociated. The dipole moment of $BCl_3 \cdot PH_3$ in benzene solution was measured as 5.2 ± 0.1 D (those for $(C_6H_5)_3PBCl_3$ and $(CH_3)_3PBCl_3$ are 7.01 ± 0.06 and 7.03 ± 0.01 D, respectively) [159].

By means of a kinetic investigation of the reaction between diborane and phosphine at 0 °C in the gas phase, Brumberger and Marcus [160] were able to describe the probable course of the reaction using the following equations:

$$B_2H_6 + PH_3 \longrightarrow (BH_3PH_3)_{(g)} + BH_3 \qquad (46)$$

$$BH_3 + PH_3 \longrightarrow (BH_3PH_3)_{(g)} \qquad (47)$$

$$(BH_3PH_3)_{(g)} \longrightarrow BH_3PH_{3\,(solid)} \qquad (48)$$

A $1:1$ adduct is formed between monobromodiborane and phosphine at –78 °C.

$$B_2H_5Br + PH_3 \xrightarrow{\ -78\,°C\ } H_3PBH_2BrBH_{3\,(solid)} \qquad (49)$$

This adduct begins to decompose to diborane and H_3PBH_2Br at –45 °C [161]. The latter compound is stable up to 0 °C. At this temperature it polymerises slowly by splitting off hydrogen:

$$H_3PBH_2Br \xrightarrow{\ 0\,°C\ } 1/n(H_2PBHBr)_n + H_2 \qquad (50)$$

When monobromodiborane and phosphine are mixed in the proportions
1 : 2 at -78 °C, equimolar amounts of H_3PBH_2Br and H_3PBH_3 are obtained
on melting the reaction mixture. Again, $H_3PBH_2BrBH_3$ is probably formed
first but, at higher temperatures, reacts further with phosphine according to
Eq. (51).

$$H_3PBH_2BrBH_{3(solid)} + PH_3 \xrightarrow{-63\,°C} H_3PBH_{3(solid)} + H_3PBH_2Br \qquad (51)$$

Recently Drake and Simpson [263] thoroughly investigated the reactions between PH_3BH_3 and PH_3BH_2Br with HCl or HBr. Phosphine-borane reacted with
equi-molar quantities of hydrogen chloride or hydrogen bromide to give the
corresponding monohaloborane adducts.

$$PH_3 \cdot BH_3 + HX \longrightarrow PH_3BH_2X + H_2 \quad (X = Cl, Br) \qquad (52)$$

With two molar amounts of the hydrogen chloride or bromide, the dihalo-
borane adducts of phosphine were formed.

$$PH_3 \cdot BH_3 + 2\,HX \longrightarrow PH_3BHX_2 + 2\,H_2 \quad (X = Cl, Br) \qquad (53)$$

Similar reactions of phosphine-monobromoborane with hydrogen halides
lead to $PH_3 \cdot BHBrCl$ and $PH_3 \cdot BHBr_2$. The n.m.r. data for the phosphine adducts are presented in Table 10.

Table 10. N.M.R. data for the adducts between phosphine and
BH_3 or haloborane (approx. 10% solutions in CH_2Cl_2 at -20 °C) [263]

Compound	$\delta_{1H}(PH_3)$[a] [ppm]	J_{PH} [Hz]	J_{HH} [Hz]
$PH_3 \cdot BH_3'$	4.24	369	8.0
$PH_3 \cdot BH_2'Cl$	4.79	389	6.2
$PH_3 \cdot BH'Cl_2$	5.14	413	3.8[b]
$PH_3 \cdot BH_2'Br$	5.08	405	6.2
$PH_3 \cdot BH'Br_2$	5.56	420	4.4
$PH_3 \cdot BH'BrCl$	5.32	417	4.0

[a] Relative to TMS internal reference.
[b] At -60 °C.

As can be seen in Table 10, the coupling constant, J_{PH}, increases with increasing Lewis acidity of the borane component. The Lewis acidity increases in
the following series of compounds thus: $BH_3 \ll BCl_3 < BBr_3$; $BH_3 < BH_2Cl <
BHCl_2$; $BH_3 < BH_2Br < BHBr_2$ and $BHCl_2 < BHBrCl < BHBr_2$. In the series,
$[PH_4]^+$, $[CH_3PH_3]^+$, $[(CH_3)_2PH_2]^+$ and $[(CH_3)_3PH]^+$, the coupling constant,
J_{PH}, decreases with increasing substitution, as must be expected when the

Fermi contact term is the most important factor for coupling. The bonding in $[PH_4]^+$ is probably comprised of pure sp^3-hybridised orbitals, so that increasing substitution by methyl groups reduces the s character of the remaining P–H bonds and simultaneously causes a reduction in the H–P–H bond angle. On the other hand, the coupling constant, J_{PH}, increases in the series PH_3, CH_3PH_2 and $(CH_3)_2PH$ with increasing methyl substitution [413]. This apparent contradiction can be explained by considering that J_{PH} is dependent upon the geometry of the molecule [414]. The replacement of hydrogen atoms by methyl groups renders the electron lone pair more accessible (see page 7). This is shown by the fact that the H–P–H bond angle in PH_3 is only 93° whereas the C–P–C bond angle in $P(CH_3)_3$ is 98°. Therefore, among other factors, an increase in the coupling constant, J_{PH}, which indicates an increase in the s character of the P–H bonds, can occur when an electron-withdrawing group is introduced.

In systems derived from $PH_3 \cdot BH_3$, changes in the coupling constant, J_{PH}, are found which are similar to those seen in compounds derived from PH_3, i.e. compounds with phosphorus in coordination number 3. The observation that the alteration does not correspond to that seen for derivatives of PH_4^+ suggests that the very weak donor-acceptor bond in $PH_3 \cdot BH_3$ does not affect the hybridisation of phosphorus, as compared with that in free PH_3, as much as might at first be expected. A small increase in the electron withdrawing action by halogen substitution of the hydrogen atoms bonded to boron causes an increase in the s character of the P–H bonds.

11. Reactions with Silanes and Alkali Aluminiumhydrides

Silyphosphine, SiH_3PH_2, can be isolated by passing equal quantities of phosphine and silane through a reaction vessel heated to 500 °C and subsequently cooling the resulting gas stream immediately to the temperature of liquid nitrogen. In a sealed reaction vessel at 450 °C and about 200 mmHg pressure other P- and Si-containing products were observed but not individually investigated [162,163].

Silylphosphine is also formed, together with disilylphosphine, $(SiH_3)_2PH$, and disilanylphosphine, $Si_2H_5PH_2$, by the application of a silent electrical discharge to a mixture of silane and phosphine. Analogous treatments of mixtures of disilane, Si_2H_6, with PH_3 or of disilylphosphine with silane, result in the formation of disilanylphosphine and silylphosphine, respectively [164,165]. Evidence for the existence of trisilylphosphine was found for the first time by Aylett, Eméleus and Maddock [166] in the reactions between silyl iodide, H_3SiI, and phosphine or between silyltrialkylammonium iodide and phosphine. According to Amberger [167] the reaction of potassium dihydrogenphosphide with silyl bromide is suitable for the preparation of trisilylphosphine. The reaction, which apparently even below $-100\,^{\circ}$C proceeds *via* many stages can be described by the empirical equation:

$$3\,KPH_2 + 3\,SiH_3Br \longrightarrow P(SiH_3)_3 + 2\,PH_3 + 3\,KBr \qquad (54)$$

Glidewell and Sheldrick [415] found that the formation of trisilylphosphine occurs according to the scheme:

$$PH_2^- + SiH_3Br \longrightarrow SiH_3PH_2 + Br^- \qquad (55)$$

$$PH_2^- + SiH_3PH_2 \longrightarrow SiH_3PH^- + PH_3 \qquad (56)$$

$$SiH_3PH^- + SiH_3Br \longrightarrow (SiH_3)_2PH + Br^- \qquad (57)$$

$$(SiH_3)_2PH + PH_2^- \longrightarrow (SiH_3)_2P^- + PH_3 \qquad (58)$$

$$(SiH_3)_2P^- + SiH_3Br \longrightarrow (SiH_3)_3P + Br^- \qquad (59)$$

The intermediate mono- and disilylphosphines can be isolated using suitable conditions. The preferential formation of trisilylphosphine is a result of the increasing acidity of the compounds in the following order, $PH_3 >$ $H_3SiPH_2 > (H_3Si)_2PH$. Trisilylphosphine reacts with water or hydrogen chloride to form disiloxane or silyl chloride, respectively. It has a much smaller nucleophilic character than trimethylphosphine [415].

At room temperature trisilylphosphine is a colourless liquid, which is spontaneously inflammable in air (extrapolated boiling point, $114\,^{\circ}$C). According to electron diffraction results the $Si-P-Si$ bond angle in trisilylphosphine is $95 \pm 2^{\circ}$ [168]. The measured $P-Si$ bond length of 2.247 ± 0.005 Å [168] corresponds to the expected $Si-P$ single bond length of 2.25 Å predicted by Beagley [169]. The ^{31}P nuclear magnetic resonance spectrum of the compound indicates that predominantly p orbitals of the phosphorus atom are used for the formation of the σ-bonding system [339]. The chemical shift of $P(SiH_3)_3$ is $+378$ ppm (relative to 85% orthophosphoric acid); this is the largest positive shift observed to date for a compound of phosphorus.

Lithium aluminiumhydride reacts with phosphine in diglyme or tetrahydrofuran to give $LiAl(PH_2)_4$ with evolution of hydrogen [170]

$$LiAlH_4 + 4\,PH_3 \longrightarrow LiAl(PH_2)_4 + 4\,H_2 \qquad (60)$$

The lithium tetrakis (dihydrogenphosphido) aluminate, which is soluble in diglyme, shows typical organic and inorganic substitution reactions and can be used to introduce PH_2 groups into molecules. The compound is quantitatively hydrolysed thus:

$$LiAl(PH_2)_4 + 4\,H_2O \longrightarrow 4\,PH_3 + LiAl(OH)_4 \qquad (61)$$

With ethyl iodide, ethylphosphine and phosphine are formed. The reaction with excess silyl bromide results in the formation of silylphosphine [416].

Silylphosphine, like methylphosphine, forms an adduct with diborane [417]

$$2\,SiH_3PH_2 + B_2H_6 \longrightarrow 2\,SiH_3PH_2 \cdot BH_3 \tag{62}$$

$$2\,CH_3PH_2 + B_2H_6 \longrightarrow 2\,CH_3PH_2 \cdot BH_3 \tag{63}$$

The proton n.m.r. spectra of these adducts have been intensively studied. The BH_3'' resonance in the spectrum of $CH_3PH_2' \cdot BH_3''$ consists of a 1 : 1 : 1 : 1 quartet due of coupling between the boron nucleus ([11]B, 80% natural abundance; $I = 3/2$) and the directly bonded protons ($J_{BH''} = 99$ Hz). Each component of the quartet is further split into a doublet of triplets due to coupling with the phosphorus nucleus ($J_{PH''} = 16$ Hz) and the two protons bonded to phosphorus, respectively. The PH_2'-signal is, as is typical for phosphine-borane adducts, a doublet with $J_{PH'} = 375$ Hz. All n.m.r. data for the two types of adducts are given in Table 11.

Table 11. ^1H N.M.R. data for silyl- and methylphosphine ($^1H_3$)- and ($^2H_3$)-borane adducts (pure liquids at -20 or $25\ ^\circ C$) [417]

Compound	Chemical shift [ppm]		
	$\delta_{(SiH)}$ or $\delta_{(CH)}$	$\delta_{(PH)}$	$\delta_{(BH)}$
$SiH_3PH_2' \cdot BH_3''$	4.0	3.8	0.8
$Si(HD)_3PH_2' \cdot B(HD'')_3$	4.0	3.8	0.7
$CH_3PH_2' \cdot BH_3''$	1.30	4.58	0.53
$CH_3PH_2' \cdot BD_3$	1.30	4.52	

Compound	Coupling constant [Hz]						
	J_{HH}	$J_{H'H''}$	J_{PH}	$J_{PH'}$	$J_{PH''}$	$J_{BH''}$	$J_{H'D}$
$SiH_3PH_2' \cdot BH_3''$	4	8	13.5	358	ca. 16	101	
$Si(HD)_3PH_2' \cdot B(HD'')_3$	4.5		14.0	255		102	
$CH_3PH_2' \cdot BH_3''$	7	7	ca. 13	375		102	
$CH_3PH_2' \cdot BD_3$	6.5		ca. 13	376			1

12. Reactions with Heavy Metals and their Ions

Heavy metal salt solutions react with phosphine in many cases to give normal phosphides. In this way, for example, Ag_3P, Au_3P, Hg_3P_2, Pb_3P_2 and Cd_3P_2 can be prepared [256,257]. The reaction of phosphine with nickel salt solutions generally results in the formation of a precipitate of composition varying from $Ni_{0.4}P$ to NiP. The nickel phosphides, Ni_5P_2, Ni_2P and NiP can only be isolated by using specific conditions [258].

No new results about the reactions of phosphine with copper salts, their aqueous solutions or with metallic copper are available. The earlier literature is often contradictory. According to Rose [457] PH_3 reacts with $CuCl_2$ to give HCl and Cu_3P_2, whereas aqueous solutions of $CuCl_2$ are not attacked by phosphine [458]. Results on the behaviour of phosphine towards copper (I) compounds have been confirmed. With acidified CuCl solution $CuCl \cdot PH_3$ is formed; $CuBr \cdot PH_3$ and $CuI \cdot PH_3$ can be prepared analogously [459,460]. Under pressure the compounds $CuCl \cdot 2 PH_3$ and $CuBr \cdot 2 PH_3$ are formed [461]. Only an earlier literature report about the reaction of PH_3 with metallic copper, in which it is reported that at 180–200 °C Cu_3P is formed, exists [458]. Also the behaviour of phosphine towards silver compounds is still unclear in many respects. On passing PH_3 into an alcoholic solution of AgI, $AgI \cdot 5 PH_3$ is obtained, while no addition compounds are formed between PH_3 and AgCl or AgBr [460]. According to a more recent report, a mixture of Ag_3P and elemental silver is obtained on treating a dilute alcoholic solution of silver nitrate with phosphine. Unstable $Ag_3P \cdot 3 AgNO_3$ is formed on passing phosphine into a concentrated aqueous solution of silver nitrate [462]. Uranium salts in aqueous solution are not attacked by PH_3 [463]. The corresponding diphosphides are formed by heating the metals Ti, Zr, V, Nb, Ta, Cr, Mo, and W with phosphine in a hydrogen atmosphere in an electric oven at 800 °C [464].

13. Reactions with Aldehydes

a) Aliphatic Aldehydes

Reactions between phosphine and aldehydes were observed relatively early by Messinger and Engels [171,172]. By passing phosphine through etheral solutions of acetaldehyde, propionaldehyde or butyraldehyde in the presence of hydrogen halides they obtained tetrakis(hydroxyethyl)-, tetrakis(hydroxypropyl)- or tetrakis(hydroxybutyl)-phosphonium halides, respectively. Tetrakis(hydroxyethyl)phosphonium iodide was isolated several years earlier by Girard [173] as the reaction product from phosphonium iodide and acetaldehyde.

Little attention was paid to these reactions until 1921 when Hoffman [174-176] prepared tetrakis(hydroxymethyl)phosphonium chloride by passing phosphine into a warm, aqueous hydrochloric acid solution of formaldehyde. The product was obtained in the form of colourless crystals by evaporation of the reaction solution.

A kinetic study of the reaction between phosphine and formaldehyde showed that it is of the first order with respect to phosphine and to the aldehyde. It is catalysed by HCl. At hydrogen chloride concentrations of less than 0.2 mol/1 the rate of reaction is proportional to the HCl-content of the solution, at higher concentrations the rate is independent of the HCl-content [177]. The dependencies found can be accounted for by both bi- and trimolecular mechanisms. In the latter case, a simultaneous interaction between one molecule of aldehyde, one of phosphine and a proton must be assumed:

$$H-\overset{\overset{\displaystyle H}{|}}{\underset{\underset{\displaystyle H}{|}}{P}}| + CH_2O + H^+ \xrightarrow{\text{slow}} H-\overset{\overset{\displaystyle H}{|}}{\underset{\underset{\displaystyle H}{|}}{P}}^+-CH_2OH \rightleftharpoons |\overset{\overset{\displaystyle H}{|}}{\underset{\underset{\displaystyle H}{|}}{P}}-CH_2OH + H^+ \quad (64)$$

The hydroxymethylphosphonium ion first formed changes into mono-hydroxymethylphosphine by releasing a proton. This phosphine reacts further in the same way as phosphine itself until finally the quarternary phosphonium ion is formed. For a bimolecular reaction mechanism, the first stage must be assumed to be the formation of a carbonium ion from the aldehyde molecule and a proton. This ion then reacts with phosphine.

$$CH_2O + H^+ \underset{\text{fast}}{\rightleftharpoons} \overset{+}{C}H_2OH \quad (65)$$

$$H-\overset{\overset{\displaystyle H}{|}}{\underset{\underset{\displaystyle H}{|}}{P}}| + \overset{+}{C}H_2OH \underset{\text{slow}}{\rightleftharpoons} H-\overset{\overset{\displaystyle H}{|}}{\underset{\underset{\displaystyle H}{|}}{P}}^+-CH_2OH \underset{\text{fast}}{\rightleftharpoons} |\overset{\overset{\displaystyle H}{|}}{\underset{\underset{\displaystyle H}{|}}{P}}-CH_2OH + H^+ \quad (66)$$

In the presence of heavy metal salts, such as $HgCl_2$, $AgNO_3$ or $PtCl_4$, the reaction between phosphine and formaldehyde is also catalysed by acids weaker than hydrochloric acid. Thus, in this way, phosphonium acetate, oxalate, lactate or fluoride can be obtained directly. According to studies by Raver [178,179] phosphine reacts with formaldehyde even in the absence of acids when metal salts such as $HgCl_2$, $NiCl_2$, $Cr_2(SO_4)_3$, $PtCl_4$ or finely divided metals are present in catalytic amounts [310,323]. Tetrakis (hydroxymethyl)phosphonium hydroxide is thus formed.

The reactions of substituted phosphines with formaldehyde show that the rate of the reaction is determined by the nucleophilic character of the phosphorus atom. The energies of activation decrease in the order [180]:

$$PH_3 > C_2H_5PH_2 \geqslant CH_3PH_2 > CH_3C_2H_5PH \geqslant (CH_3)_2PH \quad (67)$$

On the other hand, fluorine-containing substituents reduce the electron donating power of the phosphorus atom.

1,1,2,2-Tetrafluoroethylphosphine only reacts with CH_2O in the presence of hydrochloric acid to give $CHF_2CF_2P(CH_2OH)_2$ (b.p. $100-110\,°C/0.18$ mmHg, with decomposition). The corresponding phosphonium compound is not formed [181]. Phosphine reacts with aqueous solutions of formaldehyde in the presence of secondary amines to give tris(dialkylaminomethyl)phosphines in good yields [182]:

$$PH_3 + 3\,CH_2O + 3\,HNR_2 \longrightarrow (R_2NCH_2)_3P + 3\,H_2O \quad (68)$$

Secondary phosphines of the type $(CHCl_2CHOH)_2PH$, $(CCl_3CHOH)_2PH$ or $(CH_3CHClCCl_2CHOH)_2PH$ are formed by the reactions of phosphine with the corresponding aldehyde in the presence of HCl with tetrahydrofuran as reaction medium [183,184].

Tetrakis(1-hydroxyalkyl)phosphonium salts were prepared from phosphine and various aliphatic aldehydes using similar reaction conditions [185].

Chloral hydrate and chloral alcoholate react with phosphine in the presence of HCl to give compound 4 in the form of its monohydrates [333]. The preparation of this compound from chloral hydrate and phosphonium iodide has been previously described by Girard [184].

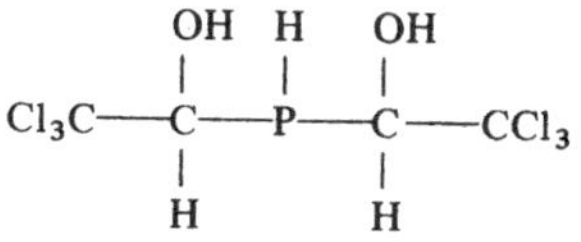

$$Cl_3C—\overset{\overset{\displaystyle OH}{|}}{\underset{\underset{\displaystyle H}{|}}{C}}—\overset{\overset{\displaystyle H}{|}}{P}—\overset{\overset{\displaystyle OH}{|}}{\underset{\underset{\displaystyle H}{|}}{C}}—CCl_3$$

4

The reactions of phosphine with α-branched aldehydes follow a different pathway. They lead to the formation of stable, heterocyclic, secondary phosphines of the following type, 5 [186]. With suitable dialdehydes, spirocyclic phosphonium salts are obtained [186,187]; these are very difficult to prepare by other methods [188]:

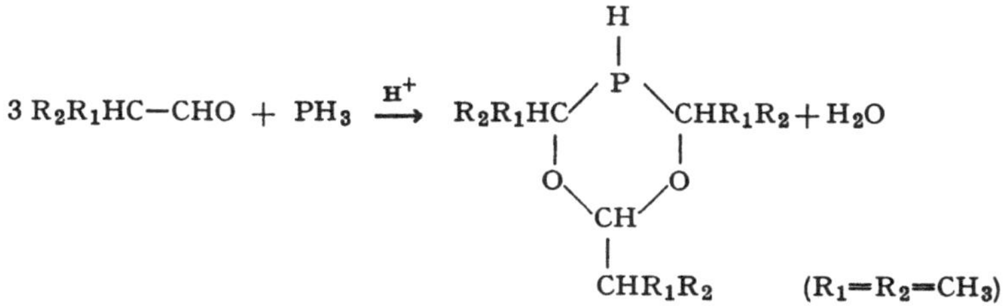

$$3\ R_2R_1HC—CHO + PH_3 \xrightarrow{H^+} \ \text{(structure 5)} + H_2O \qquad (69)$$

$$(R_1=R_2=CH_3)$$

5

$$2\ (CH_2)_n \overset{CHO}{\underset{CHO}{<}} + PH_3 + HCl \longrightarrow \left[\text{(structure)} \right] Cl \qquad (70)$$

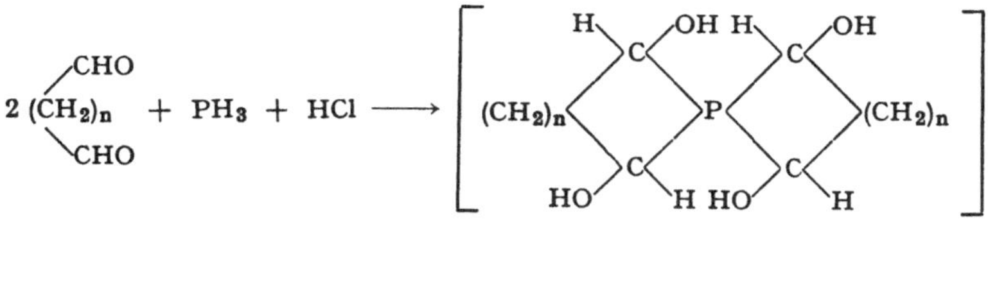

b) Aromatic Aldehydes

As previously mentioned, the first reactions between phosphine and aromatic aldehydes were carried out by Messinger and Engels [171,172], although the nature of the reaction products could only be clarified in the last few years.

E. Fluck

When a stream of phosphine is passed into an ether solution of benzaldehyde saturated with HCl, a colourless, crystalline solid, insoluble in the reaction mixture is formed. The infra-red spectrum of this product shows an absorption band at $1145\ cm^{-1}$ which is typical for the P=O group of a tertiary phosphine oxide. This, together with the chemical behaviour, indicates that the product is benzyl-bis(α-hydroxybenzyl)phosphine oxide, 6, [189].

$$(C_6H_5CHOH)_2P(O)(CH_2C_6H_5)$$

6

p-Methylbenzaldehyde and p-chlorobenzaldehyde produce the corresponding phosphine oxides in analogeous reactions [189].

When methanol is used as reaction solvent for the treatment of benzaldehyde with phosphine, however, the product is tris(α-methoxybenzyl)phosphine, 7, [190]:

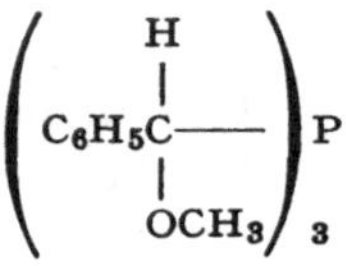

$$\left(\begin{matrix} H \\ | \\ C_6H_5C\text{——} \\ | \\ OCH_3 \end{matrix}\right)_3 P$$

7

The use of ethanol or isopropanol as solvent results in the formation of tris(α-ethoxybenzyl)- or tris(α-isopropoxybenzyl)-phosphine, respectively, in yields of up to 85% [190].

Kinetic investigations of the reactions discussed here have been made by Ettel and Horak [190].

14. Reactions with Ketones

Simple ketones react with PH_3 in strongly acid media to form primary phosphine oxides and 1-hydroxyalkyl-sec.phosphine oxides [191-193].

$$R-CO-R'+PH_3 \xrightarrow{H^+} RR'CH-PH_2(O) \tag{71}$$

$$R-CO-R'+RR'CHPH_2(O) \rightleftharpoons RR'CH-PH(O)-C(OH)RR' \tag{72}$$

The proportions of primary and secondary phosphine oxide formed are dependent mainly on steric effects.

Buckler and Epstein [192,193] suggested the following reaction mechanism to account for the formation of primary phosphine oxides:

$$R-CO-R'+PH_3 \rightleftharpoons RR'\overset{\displaystyle OH}{\underset{\displaystyle |}{C}}-PH_2 \xrightarrow[-H_2O]{H^+} RR'\overset{\oplus}{C}-PH_2 \xrightarrow{-H^+}$$

$$RR'C=PH \xrightarrow{H_2O} RR'CH-PH(OH) \rightleftharpoons RR'CH-P(O)H_2$$

(73)

The first step corresponds to a normal carbonyl addition, as is also observed with aliphatic aldehydes; but here the equilibrium does not lie so far to the right. A strongly acidic medium is necessary for the next stage, the formation of a carbonium ion. For example, it is found that PH_3 only reacts with acetone when the solution is more than 8-molar in hydrochloric acid. By analogy, the formation of a double bond between carbon and phosphorus, which in our opinion, however is improbable, is assumed to be the third stage. Finally, this is followed by the addition of water and tautomeric rearrangement to the primary phosphine. In this way, phosphine, which is present in technical acetylene, and because of its good solubility in actone concentrates in the commercial steel cylinders, forms with acetone, isopropylphosphine oxide and possibly, secondary products [418].

One of the most reactive, with respect to phosphine, ketones, hexafluorocyclobutanone, produces with phosphine primary and secondary 1-hydroxyfluorocyclobutylphosphines instead of phosphine oxides [194]:

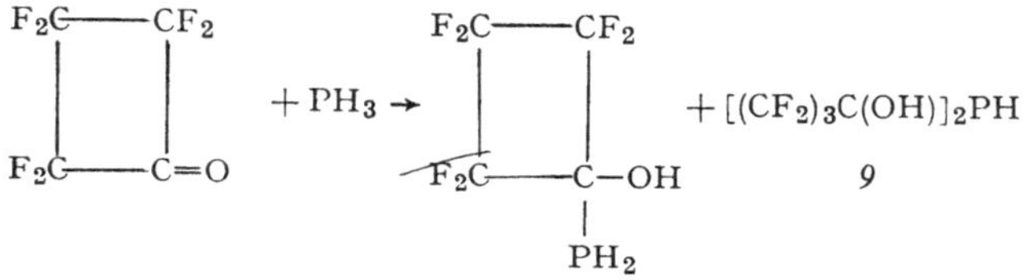

(74)

The proportions of primary and secondary phosphine formed are dependent on the molar ratio of the reaction components. Excess of phosphine leads mainly to *8*, while an excess of ketone produces a practically quantitative yield of *9*. The two phosphines *8* and *9* are both hydrolytically stable but, however, are oxidised to oily products by air. The secondary phosphine *9* reacts with diethylphenylboronate to give the ester *10* in high yield [194].

Similar to ketones, diketones also react readily with phosphine in aqueous, strong hydrochloric acid media [195]. Thus, 2,4-pentadione gives a colourless crystalline substance of composition $C_{10}H_{17}O_3P$ in good yield. On the basis of spectroscopic studies and the chemical properties of the compound the authors suggest the structure *11*.

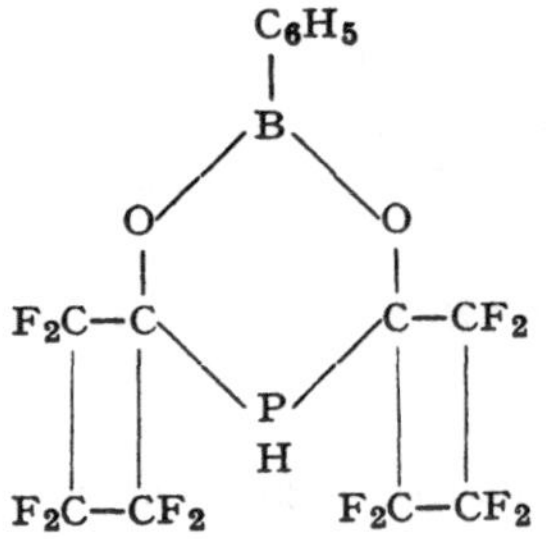

E. Fluck

10

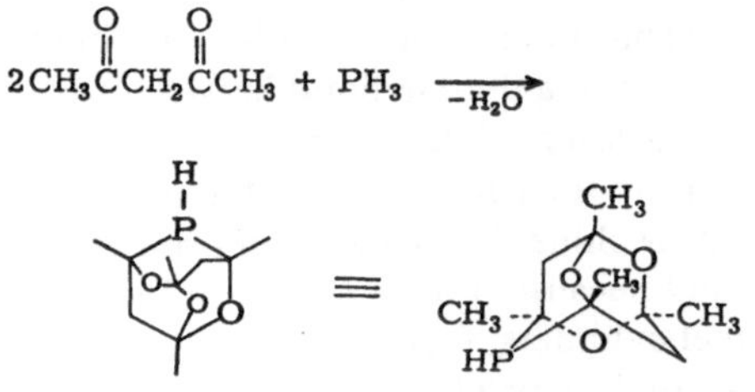

(75)

11

Analogous products are obtained from reactions between 2.4-pentadione and various primary phosphines.

Structure *12* was originally proposed for the product, with the composition $C_9H_9O_6P$, obtained by passing a mixture of phosphine and HCl into an ether solution of pyruvic acid, an α-keto acid [171,172]. However, the infra-red and nuclear magnetic resonance spectra can only be interpreted in terms of the structure *13*, recently proposed by Buckler [189]. This structure also ac-

12

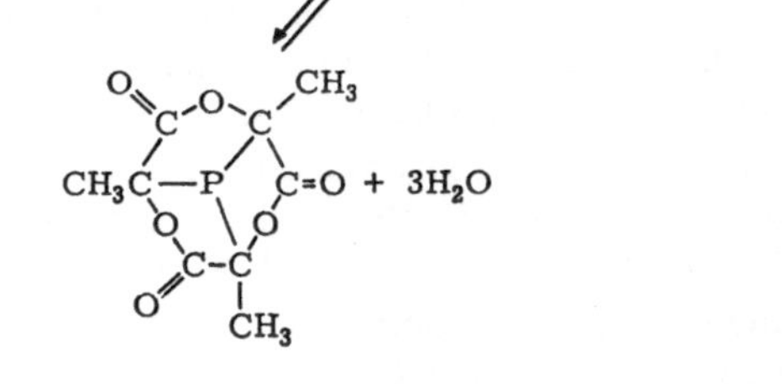

(76)

13

44

counts for the chemical behaviour of the compound. A normal carbonyl addition followed by cyclisation with the splitting off of water was assumed to account for the formation of compound *13*.

15. Reactions with Olefins

The addition of phosphine to olefins provides today a generally applicable method for the syntheses of organophosphines. Stiles, Rust and Vaughan [196] were the first to study the reaction systematically. It is catalysed by organic peroxides such as, for example, di-*t*-butyl peroxide, by α, α'-azobis-isobutyro-nitrile [197], by other free radical sources or by exposing the reaction mixture to UV- or X-radiation [197]. The PH_2 radicals, produced according to Eq. (77), react further with olefins thus producing PH_2 radicals continually.

$$PH_3 \longrightarrow PH_2\cdot + H\cdot \qquad (77)$$

On treatment of phosphine with 1-butene, cyclohexene, allyl alcohol, allylamine or allyl chloride, the corresponding primary, secondary and tertiary organophosphines are obtained in yields ranging from 2 to 67%. The reaction between phosphine and 1-butene is, among others, used for the industrial preparation of tributylphosphine [198].

The addition of phosphine to olefins is accelerated by acidic [199] and basic [200] catalysts. Under the influence of non-oxidising acids or Lewis acid such as, for example, methanesulphonic acid, benzenesulphonic acid, trifluoro-acetic acid or boron trifluoride [201] phosphine is quickly added to olefins at pressures of 20–40 at, and temperatures of 30–60 °C. It is assumed that the reaction proceeds *via* a carbonium ion which is first formed thus:

$$R_2C{=}CH_2 + H^+ \longrightarrow R_2\overset{\oplus}{C}{-}CH_3 \qquad (78)$$

This reacts further with phosphine to give the monoalkylphosphonium ion and which finally splits off a proton to form the corresponding phosphine.

$$R_2\overset{\oplus}{C}{-}CH_3 + PH_3 \longrightarrow [H_3P{-}CR_2CH_3]^+ \longrightarrow H_2P{-}CR_2CH_3 + H^+ \quad (79)$$

This assumption is supported by the fact that tertiary phosphines react especially readily under analogous conditions.

With strongly basic catalysts, in contrast, it is assumed that, in spite of the extremely weak acid character of phosphine, the reaction formally follows the Michael mechanism, *i.e.* it proceeds according to the sequence of reactions (80), (81) and (82) [202].

$$PH_3 + OH^- \longrightarrow PH_2^- + H_2O \qquad (80)$$

$$PH_2^- + CH_2CHCN \longrightarrow H_2PCH_2\bar{C}HCN \qquad (81)$$

$$H_2PCH_2\bar{C}HCN + H^+ \longrightarrow H_2PCH_2CH_2CN \qquad (82)$$

By the use of base catalysts phosphine and phenylphosphine for example can be cyanoethylated. Acrylonitrile and phosphine react together at room temperature in the presence of aqueous potassium hydroxide solution to give, depending on the reaction conditions, primary, secondary and tertiary 2-cyanoethyl-phosphines [200]:

$$PH_3 + CH_2{=}CHCN \longrightarrow H_2PCH_2CH_2CN \qquad (83)$$

$$PH_3 + 2\,CH_2{=}CHCN \longrightarrow HP(CH_2CH_2CN)_2 \qquad (84)$$

$$PH_3 + 3\,CH_2{=}CHCN \longrightarrow P(CH_2CH_2CN)_3 \qquad (85)$$

In place of potassium hydroxide, strong organic bases or suitable ion exchangers, such as, for example, Dowex-2, can also be used.

Haszeldine and co-workers investigated the reactions between phosphine and polyfluoro-olefins very carefully [203,204]. These, initiated by UV-radiation [203,204], or carried out in sealed tubes at higher temperatures, [205,206], lead to primary and secondary polyfluoroalkylphosphines. 1,1,2,2-tetrafluoroethylene, for example, forms 1,1,2,2-tetrafluoroethylphosphine in 86% yield. In addition, the reaction product also contains tetrafluoroethylenediphosphine, whereas the reaction of ethylene with phosphine produces no ethylenediphosphine. The conditions under which the reaction proceeds and the nature of the final products suggest the mechanism described by the following equations [203].

$$PH_3 \xrightarrow{\;h\cdot\nu\;} \cdot PH_2 + \cdot H \qquad (86)$$

$$\cdot PH_2 + C_2F_4 \longrightarrow H_2PCF_2CF_2\cdot \qquad (87)$$

$$H_2PCF_2CF_2\cdot + PH_3 \longrightarrow H_2PCF_2CF_2H + \cdot PH_2 \qquad (88)$$

$$H\cdot + C_2F_4 \longrightarrow CHF_2CF_2\cdot \qquad (89)$$

$$CHF_2CF_2\cdot + PH_3 \longrightarrow CHF_2CF_2H + PH_2\cdot \longrightarrow etc. \qquad (90)$$

$$H_2PCF_2CF_2\cdot + \cdot PH_2 \longrightarrow H_2PCF_2CF_2PH_2 \qquad (91)$$

$$CHF_2CF_2PH_2 \xrightarrow{\;h\cdot\nu\;} CHF_2CF_2\dot{P}H \xrightarrow[PH_3]{C_2F_4} (CHF_2CF_2)_2PH \qquad (92)$$

An equi-molar mixture of chlorotrifluoroethylene and phosphine reacts in UV light to give a 91% yield of chlorotrifluoroethylphosphine, $ClCHFCF_2PH_2$, i.e. phosphine attacks the CF_2 group, as is known from other radicals [203]. In contrast, 1,1-difluoroethylene with phosphine gives $CHF_2CH_2PH_2$ as main product as well as the secondary phosphine $(CHF_2CH_2)_2PH$, *i.e.* here the attack occurs only at the CH_2 group of 1,1-difluoroethylene. Surprisingly $FClC=CFCl$ does not react with phosphine. Finally the photochemical reaction of phosphine with trifluoroethylene results in the formation of $CF_3CHFCF_2PH_2$ and $CF_3CF(PH_2)CHF_2$ in the ratio 85 : 15.

The addition of phosphine to 5,6-dideoxy-1,2-O-isopropylidene-D-xylo-hex-5-enfuranose (*14*) takes place when the reaction mixture is irradiated with UV-light. A mixture of 5,6-dideoxy-1,2-O-isopropylidene-α-D-xylo-hexa-furanose-phosphine (*15*) and bis-6-(5,6-dideoxy-1,2-O-isopropylidene-α-D-xylo hexanose)phosphine (*16*) is probably formed but the components could not be separated. In the presence of atmospheric oxygen these are converted to the corresponding phosphonous acid (*17*) and the secondary phosphine oxide (*18*), respectively [419].

$$(93) - (96)$$

16. Reactions with Isocyanates

Phosphine reacts with aryl isocyanates to form tricarbamoylphosphines (*19*) [207,208]:

$$PH_3 + 3 \; X-\!\!\!\left\langle\!\!\bigcirc\!\!\right\rangle\!\!-NCO \;\xrightarrow{N(C_2H_5)_3}\; P\left[C(O)NH-\!\!\!\left\langle\!\!\bigcirc\!\!\right\rangle\!\!-X\right]_3 \qquad (97)$$

19

$$X = H, Cl, NO_2$$

The yields from the reactions, carried out at room temperature and a pressure of 2–4 at, increase with increasing electronegativity of the substituent X. The yield with phenyl isocyanate is 13%, with *p*-chlorophenyl isocyanate 55% and with *p*-nitrophenyl isocyanate it is 100%. Primary and secondary carbamoylphosphines cannot be isolated even when equi-molar quantities of phosphine and isocyanate are used. Their intermediate formation is probable but apparently they are more reactive towards isocyanates than phosphine itself. Similarly, phosphine does not react with free cyanic acid whereas primary and secondary phosphines react with cyanic acid, as with isocyanates, to form the corresponding carbamoylphosphines [209]. Attempts to make phosphine react with phenyl isothiocyanate did not succeed [210].

17. Reactions with Aromatic Acid Chlorides

Aromatic acid chlorides react with phosphine at 50 °C in absolute pyridine to form mono-, di- and triacylphosphines. For example, PH_3 and benzoyl chloride give tribenzoylphosphine, a yellow crystalline compound which is resistent to water and dilute acids but is hydrolysed to PH_3 and alkali benzoate by alkalis [217,218].

18. Reactions with Trimethylindium

When phosphine is passed through a benzene solution of trimethylindium at 0–25 °C, an exothermic reaction occurs and a pale yellow precipitate is formed. This is probably $(CH_3In \cdot PH)_n$; it is insoluble in the usual organic solvents and is spontaneously inflammable in air.

In water it decomposes to methane and phosphine. At 100–120 °C the substance sometimes decomposes with evolution of smoke or explosively. On heating to 250–270 °C, indium phosphide is formed as the residue [212].

Trimethylindium is readily soluble in liquid phosphine. A 1 : 1 adduct, $(CH_3)_3In \cdot PH_3$, is probably formed at −123 °C. Above this temperature, at about $-112\,°C$, it decomposes to $(CH_3)_3In$ and phosphine. At still higher temperatures, above about $-78\,°C$, condensation of the monomeric molecules occurs with the evolution of methane. The polymer of empirical composition $CH_3In \cdot PH$ thus formed, is stable up to about $95\,°C$ and decomposes at higher temperatures to CH_4 and InP [213].

19. Phosphine as Ligand in Coordination Compounds

A deep-blue solution of $V(CO)_4PH_2$ is formed on passing PH_3 into a hexane solution of vanadium hexacarbonyl at room temperature [211]. A molecular weight determination of a benzene solution of the diamagnetic compound gave a value of twice the formula weight. Thus, apparently, a dimeric complex is formed. The infra-red spectrum confirms that PH_2 bridging groups with σ-bonds between vanadium and phosphorus are present.

$$2\ V(CO)_6 + 2\ PH_3 \xrightarrow{\text{Hexane}} (CO)_4V \begin{array}{c} \diagup PH_2 \diagdown \\ \diagdown PH_2 \diagup \end{array} V(CO)_4 + 4\ CO + H_2 \qquad (98)$$

20

Di-μ-phosphino-bis(tetracarbonylvanadium), *20*, is very soluble in hexane, benzene and methylene chloride and can be kept under a nitrogen atmosphere.

Mono- and diphenylphosphine also give the corresponding dimeric complexes with vanadium hexacarbonyl.

Whereas the above described reactions result only in dimeric compounds of the type *20*, E. O. Fischer and co-workers [335] were recently able to prepare a carbonyl compound in which phosphine acts as a mono-dentate ligand. Tricarbonylcyclopentadienylphosphine vanadium, π-$C_5H_5V(CO)_3PH_3$, *21*, was obtained by exposing a solution of π-$C_5H_5V(CO)_4$ in tetrahydrofuran or benzene under a phosphine atmosphere to UV light for several hours according to Eq. (99).

$$\pi\text{-}C_5H_5V(CO)_4 + PH_3 \xrightarrow{h\cdot\nu} \pi\text{-}C_5H_5V(CO)_3PH_3 + CO \qquad (99)$$

21

The ^1H n.m.r. spectrum shows a doublet with a chemical shift of $\tau = 6.81$ and a coupling constant J_{P-H} of 324 Hz.

The following compounds can be obtained from analogous reactions:

$C_5H_5Mn(CO)_2PH_3$ (red-brown, m.p. $72\,°C$; $\tau_{PH_3} = 6.76$; $J_{P-H} = 327$ Hz)

$Cr(CO)_5PH_3$ (pale yellow, m.p. $116\,°C$, $\tau_{PH_3} = 7.55$; $J_{P-H} = 337.5$ Hz)

$Fe(CO)_4PH_3$ (bright yellow, m.p. $36\,°C$, $\tau_{PH_3} = 7.85$; $J_{P-H} = 365$ Hz)

$Mo(CO)_5PH_3$ (colourless, m.p. $112\,°C$ (decomp.) $\tau_{PH_3} = 7.69$; $J_{P-H} = 327$ Hz)

$W(CO)_5PH_3$ (colourless, m.p. $120\,°C$, $\tau_{PH_3} = 7.49$; $J_{P-H} = 341$ Hz)

Phosphine can also replace a carbonyl group in bromopentacarbonyl manganese [336]. The compound, $BrMn(CO)_4PH_3$, shows a doublet in the ^1H n.m.r. spectrum, $\tau = 6.58$, $J_{P-H} = 355$ Hz.

PF_3 as well as CO, as ligand in coordination compounds, can also be partially replaced by PH_3 [337]. $HCo(PF_3)_3PH_3$, a light yellow, sublimable compound (m.p. 25 °C) is obtained when a mixture of $HCo(PF_3)_4$ and PH_3 is exposed to sunlight or to UV radiation from a mercury lamp, or when $HCo(PF_3)_3CO$ is treated with PH_3. The ^1H n.m.r. spectrum of liquid $HCo(PF_3)_3PH_3$ shows a doublet of quartets, $\tau = 6.12$, $J_{P-H} = 352$, $J_{F_3P-H} = 17.6$ Hz, for the PH_3 group and a broad signal at $\tau = 24.4$ for the hydrogen atom bonded to cobalt. The hydrogen atom and the PH_3 group probably occupy axial positions of the trigonal bipyramid and the PF_3 groups the equatorial positions.

Klanberg and Muetterties [336] described the introduction of two phosphine ligands into a carbonyl compound. They reacted the octahydrotriboro-tetracarbonyl-metal anions (22), where M = Cr, Mo or W, with phosphine and obtained the sublimable bisphosphino-metal tetracarbonyls (23) which are stable in air:

$$(OC)_4MB_3H_8^- + 2\ PH_3 \longrightarrow (OC)_4M(PH_3)_2 + B_3H_8^- \qquad (100)$$

$$22 \qquad\qquad\qquad 23$$

$$(M = Cr, Mo, W)$$

The chemical shifts in the ^1H and ^{31}P n.m.r. spectra of the molybdenum and tungsten compounds are $\tau = 6.31$ and 5.98 and $\delta_P = 155$ and 175 ppm (relative to H_3PO_4), respectively. The coupling constants, are 324 and 338 Hz, respectively.

Recently a compound with three PH_3 ligands bonded to a central atom was reported. Phosphine reacted rapidly and quantitatively at room temperature with a solution of tricarbonyl-hexamethylborazine chromium(0) in cyclohexane to give the octahedral tricarbonyltris(phosphine)chromium(0) in the *cis* configuration [340].

The structure of a series of other coordination compounds containing the PH_3 ligand, the preparations of which were described by Klanberg and Muetterties [336] according to the following Eqs. (101) to (107), are not individually known.

$$PH_3 + [(C_6H_5)_3P]_2PdCl_2 \longrightarrow \{[(C_6H_5)_3P](PH_3)PdCl\}_4 \qquad (101)$$

$$[(C_6H_5)_3P]_2PtI_2 \longrightarrow Pt_3[P(C_6H_5)_3]_3(PH_3)_3I_2 \qquad (102)$$

$$Ru(CO)_2Cl_2 \longrightarrow Ru_3(CO)_8(PH_3)_4 \qquad (103)$$

$$Rh_2(CO)_4Cl_2 \longrightarrow Rh_6(CO)_8(PH_3)_8 \qquad (104)$$

$$\text{Ni}(C_5H_5)_2 \longrightarrow \text{Ni}(C_5H_5)_2(PH_3)_2 \qquad (105)$$

$$[(C_6H_5)_3P]_3RhCl \longrightarrow [(C_6H_5)_3P]_2Rh(PH_3)Cl \qquad (106)$$

$$[(C_6H_5)_3P]_2Ir(CO)Cl \longrightarrow [(C_6H_5)_3P](PH_3)Ir(CO)Cl \qquad (107)$$

The authors strongly emphasise the facility with which metal clusters are apparently formed by these reactions.

V. Diphosphine and Higher Phosphines

Diphosphine, P_2H_4, is a liquid at room temperature. The vapour pressure at 0 °C is 73.0 mmHg, at –33.5 °C it is 10 mmHg [420]. The average values for the extrapolated boiling point and the latent heat of evaporation of diphosphine are 63.5 °C and 6889 cal/mol, respectively [420-422]. The results are calculated from the values reported in the literature for the vapour pressure at various temperatures; however, some of these values vary greatly from one another. The melting point has been reported to be –99 °C [421]. The heat of formation, calculated from the heat of explosion, for the formation of P_2H_4 from the elements was reported to be H^0_{298} = 5 ± 1.0 kcal/mol [423]. The most recent determination of the density of liquid diphosphine has shown that it is less dense than water [424]. Earlier, the density of liquid phosphine at 16 °C was reported to be 1.016 g/cm^3 [422]. Solid diphosphine has a density of 0.9 g/cm^3 [424]. At –136 °C the lattice is built of monoclinic, and possibly also rhombic, unit cells each containing 2 molecules (space group C_2^1 or C_s^1) [424].

The point group of the molecule is probably C_2 [424-426]. The distances between the nuclei, calculated from the force constants, are $d(P–P) = 2.11$ Å and $d(P–H) = 1.44$ Å [425]. The IR and Raman spectra of diphosphine were carefully studied by Baudler and Schmidt [425] and by Nixon [424] and the n.m.r spectra by Lynden-Bell [427].

Diphosphine is formed by the hydrolysis of calcium phosphide [422,424,425, 428,429] and also by the hydrolysis of other phosphides when these contain P–P linkages. Thus, for example, it is reported that the phosphine obtained by the hydrolysis of aluminium phosphide, which has been prepared from the elements with phosphorus in excess, is spontaneously inflammable. This is caused by the diphosphine formed at the same time. When the aluminium phosphide is prepared using stoichiometric or even excess amounts of aluminium, the formation of diphosphine is not observed on hydrolysis. The diphosphine, formed in large quantities by the hydrolysis of calcium phosphide, can be separated from the phosphine and hydrogen evolved simultaneously by cooling the gas mixture to $-78\ °C$ and diphosphine can be subsequently purified by distillation under high vacuum [425]. Diphosphine can also be obtained from the reaction of pure white phosphorus with potassium hydroxide solution at $60\ °C$ [421]. The formation of diphosphine has also been observed in the following reactions: by heating red phosphorus in a stream of hydrogen [430], by heating a mixture of red phosphorus and $Ba(OH)_2$ [431], by the action of atomic hydrogen on red phosphorus [432] and by the reaction of H_3PO_4 with acetyl chloride [433].

With B_2H_6, diphosphine forms an adduct, $P_2H_4 \cdot B_2H_6$; with boron trifluoride at low temperatures the adduct $P_2H_4 \cdot 2BF_3$ is formed [434]. On heating or exposure to light, diphosphine decomposes to PH_3 and higher phosphorus-hydrogen compounds [420,421,422,428,435].

Triphosphine, P_3H_5, has been identified as a product from the hydrolysis of Ca_3P_2 or Mg_3P_2 in acidic media by mass spectroscopy [438]. In addition, using mass spectroscopic analysis Baudler and her co-workers [436,437] were able to identify the following higher phosphines formed during the hydrolysis or the disproportionation of diphosphine:

	P_nH_{2n+2}	P_nH_n	P_nH_{n-2}	P_nH_{n-4}	Others
Triphosphine	P_3H_5[a]	P_3H_3[a]			
Tetraphosphine	P_4H_6[a]	P_4H_4[a]	P_4H_2		
Pentaphosphine	P_5H_7	P_5H_5[b]	P_5H_3[b]		
Hexaphosphine	P_6H_8	P_6H_6	P_6H_4[b]	P_6H_2	
Heptaphosphine	P_7H_9	P_7H_7	P_7H_5	P_7H_3[b]	
Octaphosphine		P_8H_8	P_8H_6	P_8H_4[b]	
Nonaphosphine		P_9H_9		P_9H_5	P_9H_3[b]
Decaphosphine		$P_{10}H_{10}$	$P_{10}H_8$		$P_{10}H_4, P_{10}H_2$[b]
Undecaphosphine					$P_{11}H_6, P_{11}H_3$[b]
Dodecaphosphine					$P_{12}H_4$
Tetradecaphosphine					$P_{14}H_{(?)}$

[a] Main products.
[b] Predominant products for the particular value of n.

The spontaneously inflammable nature of the higher phosphines decreases with increasing phosphorus content. At room temperature or on exposure to light, phosphorus-rich, yellow, solid phosphines are rapidly formed; these can also be obtained directly by thermal decomposition of diphosphine. The literature on these types of higher phosphorus hydrides which are, in general, solid and are thus refered to as "solid phosphorus hydrides" is abundant and often contradictory [439]. These are probably not discreet compounds but highly polymeric substances with no stoichiometric composition. They are probably composed of the following structural units:

$$
\begin{array}{ccc}
\mathrm{H}\!\!\diagdown \\[-4pt]
\qquad\mathrm{P}\!\!-\qquad & -\mathrm{P}- & \cdot-\mathrm{P}- \\[-4pt]
\mathrm{H}\!\!\diagup \\
& \mathrm{H} &
\end{array}
$$

| End group | Middle group | Branching group |

The solid phosphorus hydrides are insoluble in all solvents generally used. They are relatively stable in air but are oxidised by strong oxidising agents.

Translated from the German by Dr. R. E. Dunmur

VI. Literature

[1] Gengembre, P.: Hist. Mém. Acad. Roy. Soc. *10*, 651 (1785).

[2] Gengembre, P.: Ann. Crell *1789*, I, 450.

[3] Kirwan, R.: Phil. Trans. Roy. Soc. (London) *76*, 11 (1786).

[4] Gay-Lussac, L. J., Thénard, L. J.: Rech. Phys.-Chim. *1*, 184 (1811).

[5] Davy, H.: Schweiggers J. Chem. Phys. *1*, 473 (1811); *7*, 494 (1813).

[6] Rossini, F. D., Wagman, D. D., Evans, W. H., Levine, S., Jaffé, I.: Natl. Bur. Std. (U.S.), Circ. No. 500,571 (1952).

[7] Clusius, K., Frank, A.: Z. Physik. Chem. *B 34*, 405 (1936).

[8] Stephenson, C. C., Giauque, W. F.: J. Chem. Phys. *5*, 149 (1937).

[9] Clusius, K., Weigand, K.: Z. Physik. Chem. *B 46*, 1 (1940).

[10] Hardin, A. H., Harvey, K. B.: Can. J. Chem. *42*, 84 (1964).

[11] Ritchie, M.: Proc. Soc. (London) *A 128*, 551 (1930).

[12] Moles, E.: Bull. Soc. Chim. Belges *62*, 67 (1953).

[13] Natta, G., Casazza, E.: Gazz. Chim. Ital. *60*, 851 (1930).

[14] Ter Gazarian, G.: J Chim. Phys. *7*, 337 (1909), *9*, 101 (1911); Compt. Rend. *148*, 1397 (1909).

[15] Stock, A., Henning, F., Kuss, E.: Ber. Deut. Chem. Ges. *54 B*, 1119 (1921).

[16] Skinner, S.: Proc. Roy. Soc. (London) *42*, 283 (1887).

[17] Frank, A., Clusius, K.: Z. Physik. Chem. *B 42*, 395 (1939).

[18] Herzberg, G.: The structure of diatomic molecules. New York: Van Nostrand 1950.

[19] McConaghie, V. M.: Proc. Natl. Acad. Sci. U.S. *34*, 455 (1948).

[20] McConaghie, V. M., Nielsen, H. H.: J. Chem. Phys. *21*, 1836 (1953).

[21] Nielsen, H. H.: J. Chem. Phys. *20*, 759 (1952).

[22] Weston, R. E., Sirvetz, M. H.: J. Chem. Phys. *20*, 1820 (1952).

[23] Morse, P. M., Stückelberg, E. C. G.: Helv. Phys. Acta *4*, 337 (1931).

[24] Sundaram, S., Suszek, F., Cleveland, F. F.: J. Chem. Phys. *32*, 251 (1960).

[25] Siebert, H.: Z. Anorg. Allgem. Chem. *274*, 24 (1953).

[26] Yost, M., Anderson, T. F.: J. Chem. Phys. *2*, 624 (1934).

E. Fluck

27) Burrus, C. A., Jacke, A., Gordy, W.: Phys. Rev. *95*, 706 (1954).
28) Loomis, C. C., Strandberg, M. W. P.: Phys. Rev. *81*, 798 (1951).
29) Sirvetz, M. H., Weston, R. E.: J. Chem. Phys. *21*, 898 (1953).
30) Van Wazer, J. R., Callis, C. F., Shoolery, J. N., Jones, R. C.: J. Am. Chem. Soc. *78*, 5715 (1956).
31) Bartell, L. S.: J. Chem. Phys. *32*, 832 (1960).
32) Bartell, L. S., Brockway, L. O.: J. Chem. Phys. *32*, 512 (1960).
33) Bartell, L. S., Hirst, R. C.: J. Chem. Phys. *31*, 449 (1959).
34) Kuchitsu, K.: J. Mol. Spectry. *7*, 399 (1961).
35) Lippincott, E. R., Dayhoff, M. O.: Spectrochim. Acta *16*, 807 (1960).
36) Moccia, R.: J. Chem. Phys. *37*, 910 (1962).
37) Kojima, T., Breig, E. L., Lin, C. C.: J. Chem. Phys. *35*, 2139 (1960).
38) Pratt, L., Richards, R. E.: Trans. Faraday Soc. *50*, 670 (1954).
39) Van Wazer, J. R.: Phosphorus and its compounds, Vol. I. New York: Interscience Publ. 1958.
40) Cottrell, T. L.: The strength of chemical bonds, p. 271. London: Butterworth's 1958.
41) Banyard, K. E., Hake, R. B.: J. Chem. Phys. *43*, 2684 (1965).
42) Burrus, C. A.: J. Chem. Phys. *28*, 427 (1958).
43) Weaver, J. R., Parry, R. W.: Inorg. Chem. *5*, 718 (1966).
44) Gibbs, J. H.: J. Phys. Chem. *59*, 644 (1955).
45) Gibbs, J. H.: J. Chem. Phys. *22*, 1460 (1954).
46) Wilmshurst, J. K.: J. Chem. Phys. *33*, 813 (1960).
47) Van Wazer, J. R.: J. Am. Chem. Soc. *78*, 5709 (1956).
48) Paddock, N. L.: Structure and reactions in phosphorus chemistry, The Royal Institute of Chemistry, Lecture Series 1962, No. 2, p. 4.
49) Lynden-Bell, R. M.: Trans. Faraday Soc. *57*, 888 (1961).
50) Gillespie, R. J.: J. Am. Chem. Soc. *82*, 5978 (1960).
51) Mulliken, R. S.: J. Am. Chem. Soc. *77*, 887 (1955).
52) Hutchinson, D. A.: Can. J. Chem. *44*, 2711 (1966).
53) Fung, L. W., Barker, E. F.: Phys. Rev. *45*, 238 (1934).
54) Howard, J. B.: J. Chem. Phys. *3*, 207 (1935).
55) Lee, E., Wu, C. K.: Trans. Faraday Soc. *35*, 1366 (1939).
56) McKean, D. C., Schatz, P. N.: J. Chem. Phys. *24*, 316 (1956).
57) Melville, H. W.: Nature *129*, 546 (1932).
58) Cheesman, G. H., Emeléus, H. J.: J. Chem. Soc. *1932*, 2847.
59) Mayor, L., Walsh, A. D., Warsop, P.: J. Mol. Spectry. *10*, 320 (1963).
60) Halmann, M.: J. Chem. Soc. *1963*, 2853.
61) Stevenson, D. P., Coppinger, G. M., Forbes, J. W.: J. Am. Chem. Soc. *83*, 4350 (1961).
62) Walsh, A. D., Warsop, P. A.: Advan. Mol. Spectry. *2*, 582 (1962).
63) Wada, Y., Kiser, R. W.: Inorg. Chem. *3*, 174 (1964).
64) Neuert, H., Clasen, H.: Z. Naturforsch. *7a*, 410 (1952).
65) Saalfeld, F. E., Svec, H. J.: Inorg. Chem. *2*. 46 (1963).
66) Saalfeld, F. E., Svec, H. J. Inorg. Nucl. Chem. *18*, 98 (1961).
67) Watanabe, K.: J. Chem. Phys. *26*, 542, 1773 (1957).
68) Frost, D. C., McDowell, C. A.: Can. J. Chem. *36*, 39 (1958).
69) Varsel, C. J., Morrell, F. A., Resnik, F. E., Powell, W. A.: Anal. Chem. *32*, 182 (1960).
70) Dibeler, V. H., Franklin, J. L., Reese, R. M.: J. Am. Chem. Soc. *81*, 68 (1959).
71) Kley, D., Welge, K. H.: Z. Naturforsch. *20a*, 124 (1965).
72) Norrish, R. G. W., Oldershaw, G. A.: Proc. Roy. Soc. (London) *A 262*, 1 (1961).
73) Weston, R. E.: J. Am. Chem. Soc. *76*, 1027 (1954).
74) Weston, R. E., Bigeleisen, J.: J. Am. Chem. Soc. *76*, 3074 (1954).

75) Sheldrick, G. M.: Trans. Faraday Soc. *1967*, 1077.
76) Wendlandt, W.: Science *122*, 831 (1955).
77) Waddington, F. C.: Trans. Faraday Soc. *61*, 2652 (1965).
78) Sheldon, J. C., Tyree, S. Y.: J. Am. Chem. Soc. *80*, 2117 (1958).
79) Thomson, T.: Ann. Phil. Thomson [2] *8*, 87 (1816).
80) Pearson, G.: Phil. Trans. Roy. Soc. (London) *1792*, 289.
81) Thénard, P.: Compt. Rend. *18*, 652 (1844); *19*, 313 (1844); Ann. Chim. Phys. [3], *14*, 5 (1845).
82) Moissan, H.: Compt. Rend. *128*, 787 (1899).
83) Dalton, J.: Ann. Phil. Thomson [2]*11*, 7 (1818).
84) Schwarz, H.: Dinglers Polytech. J. *191*, 396 (1869).
85) Brandstätter, F.: Z. Phys. Chem. Unterricht *11*, 65 (1898).
86) Lüpke, R.: Z. Phys. Chem. Unterricht *3*, 280 (1890).
87) Saalfeld, F. E., Svec, H. J.: IS-386, 68 (1961); C.A. *57*, 236 (1962).
88) Fonzes-Diacon, H.: Compt. Rend. *130*, 1314 (1900).
89) Quesnel, G.: Compt. Rend. *253*, 1450 (1961).
90) Bodroux, F.: Bull. Soc. Chim. France [3] *27*, 568 (1902).
91) Kuznetsov, E. V., Valetdinov, R. K., Zavlina, P. M.: USSR Pat. 125 551 (1960).
92) Kuznetsov, E. V., Valetdinov, R. K., Roitburd, T. Ya., Zakharova, L. B.: Tr. Kazakhsk. Khim.-Tekhnol. Inst. *1960*, 20; C.A. *58*, 547 (1963).
93) Pass, F., Steininger, E., Zorn, H.: Monatsh. Chem. *93*, 230 (1962).
94) Baudler, M., Ständeke, H., Borgardt, M., Strabel, H., Dobbers, J.: Naturwissenschaften *53*, 106 (1966).
95) Fluck, E., Novobilsky, V.: Unpublished results.
96) Landolt, H.: Liebigs Ann. Chem. *116*, 193 (1860).
97) Matignon, C.: Compt. Rend. *130*, 1391 (1900).
98) Moser, L., Brukl, A.: Z. Anorg. Allgem. Chem. *121*, 73 (1922).
99) White, W. E., Bushey, A. H.: J. Am. Chem. Soc. *66*, 1666 (1944).
100) Montignie, E.: Bull. Soc. Chim. France *1946*, 276.
101) Dumas, J. B. A.: Ann. Chim. Phys. [2]*31*, 113 (1826).
102) Rose, H.: Ann. Physik [2] *24*, 109, 295 (1832); *32*, 467 (1834).
103) Weyl, T.: Ber. Deut. Chem. Ges. *39*, 1307 (1906).
104) Commaille, A.: J. Pharm. [2] *8*, 321 (1868).
105) Hofmann, A. W.: Ber. Deut. Chem. Ges. *4*, 200 (1871).
106) Wartik, T., Apple, E. F.: J. Am. Chem. Soc. *80*, 6155 (1958).
107) Rose, H.: Ann. Physik [2] *8*, 191 (1826); *24*, 109 (1832).
108) Martin, D. R., Dial, R. E.: J. Am. Chem. Soc. *72*, 852 (1950).
109) Houton de la Billardière, J. J.: Ann. Chim. Phys. *6*, 304 (1817).
110) Lepsius, B.: Ber. Deut. Chem. Ges. *23*, 1642 (1890).
111) Rammelsberg, C.: Ber. Deut. Chem. Ges. *6*, 88 (1873).
112) Messinger, J., Engels, C.: Ber. Deut. Chem. Ges. *21*, 326 (1888).
113) Stock, A., Henning, F., Kuss, E.: Ber. Deut. Chem. Ges. *54 B*, 1119 (1921).
114) Robertson, R., Fox, J. J., Hiscocks, E. S.: Proc. Roy. Soc. (London) *A 120*, 149 (1928).
115) Paddock, N. L.: Nature *167*, 1070 (1951).
116) Gunn, S. R., Green, L. G.: J. Phys. Chem. *65*, 779 (1961).
117) Wiberg, E., Müller-Schiedmayer, G.: Chem. Ber. *92*, 2372 (1959).
118) Niederl. Pat.-Anm. 6 504 634 (1965); C.A. *64*, 13803 (1966).
119) Palmer, M. G.: Brit. Pat. 943 281 (1963); C.A. *60*, 6524 (1964).
120) Baudler, M., Schellenberg, D.: Z. Anorg. Allgem. Chem. *340*, 113 (1965).
121) Matignon, C., Trannoy, R.: Compt. Rend *148*, 167 (1909).

E. Fluck

122) Royen, P., Hill, K.: Z. Anorg. Allgem. Chem. *229*, 112 (1936).
123) Addison, W. E., Plummer, J.: Chem. Ind. (London) *1961*, 935.
124) Joannis, A.: Ann. Chim. Phys. [8] *7*, 101 (1906).
125) Joannis, A.: Compt. Rend. *119*, 557 (1894).
126) Royen, P., Zschaage, W., Wutschel, A.: Angew. Chem. *67*, 75 (1955).
127) Knunyants, I. L., Sterlin, R. N.: Dokl. Akad. Nauk SSSR *56*, 49 (1947); C.A. *42*, 519 (1948).
128) Wagner, R. I., Burg, A. B.: J. Am. Chem. Soc. *75*, 3869 (1953).
129) Legoux, C.: Compt. Rend. *207*, 634 (1938).
130) Teach, E. G., Leffler, A. J.: USP 2964379.
131) Albers, H., Schuler, W.: Ber. Deut. Chem. Ges. *76*, 23 (1943).
132) Bergerhoff, G., Schultze-Rhonhoff, E.: Acta Cryst. *15*, 420 (1962).
133) Knoll, F., Bergerhoff, G.: Monatsh. Chem. *97*, 808 (1966).
134) Brauer, G., Zintl, E.: Z. Physik. Chem. *37 B*, 323 (1937).
135) Endrzheevskaya, S. N., Samsonov, G. V.: Zh. Obshch. Khim. *35*, 1983 (1965); C.A., *64*, 6064 (1966).
136) Samsonov, G. V., Vereikina, L. L., Endrzheevskaya, S. N., Tikhonova, N. N.: Ukr. Khim. Zh. *32*, 115 (1966); C.A. *64*, 15348 (1966).
137) Vickery, R. C.: Nature *184*, 268 (1959).
138) Williams, F. V., Ruehrwein, R. A.: J. Am. Chem. Soc. *82*, 1330 (1960).
139) Effer, D., Antell, G. R.: J. Electrochem. Soc. *107*, 252 (1960).
140) Samsonov, G. V., Vereikina, L. L., Titkov, Yu. V.: Zh. Neorgan. Khim *6*, 749 (1961); C.A. *56*, 15 125 (1962); Pat. USSR 136327 (1961); *55*, 21511 (1961).
141) Tikhonova, N. N.: Azerb. Khim. Zh. *1965*, 139; C.A. *64*, 18935 (1966).
142) Vereikina, L. L., Samsonov, G. V.: Zh. Neorgan. Khim. *5*, 1888 (1960); C.A. *57*, 1834 (1962).
143) Samsonov, G. V., Vereikina, L. L.: USSR Pat. 127 028 (1960); C.A. *54*, 18912 (1960).
144) Sheldrick, G. M.: Trans.Faraday Soc. *1967*, 1065.
145) Ebsworth, E. A. V., Sheldrick, G. M.: Trans.Faraday Soc. *1967*, 1071.
146) Fischer, H., Wiberg, E.: DBP 1042539 (1958); C.A. *54*, 20519 (1960); C.Z. *1959*, 6593.
147) Vickery, R. C.: Nature *184*, 268 (1959).
148) Schumann, H., Schwabe, P., Schmidt, M.: Inorg. Nucl. Chem. Letters *2*, 309 (1966).
149) Schumann, H., Roth, A., Stelzer, O., Schmidt, M.: Inorg. Nucl. Chem. Letters *2*, 311 (1966).
150) Parshall, G. W., Lindsey, R. V.: J. Am. Chem. Soc. *81*, 6273 (1959).
151) Bürger, H.: Personal communication.
152) Stock, A., Kuss, E.: Ber. Deut. Chem. Ges. *56 B*, 789 (1923).
153) Stock, A., Pohland, E.: Ber. Deut. Chem. Ges. *59 B*, 2215 (1926).
154) Stock, A., Wiberg, E., Martini, H., Nicklas, A.: Ber. Deut. Chem. Ges. *65 B*, 1711 (1932).
155) Gamble, E. L., Gilmont, P.: J. Am. Chem. Soc. *62*, 717 (1940).
156) Rudolph, R. W., Parry, R. W., Farran, C. F.: Inorg. Chem. *5*, 723 (1966).
157) Besson, A.: Compt. Rend. *110*, 516 (1890).
158) Tierney, P. A., Lewis, D. W., Berg, D.: J. Inorg. Nucl. Chem. *24*, 1163 (1962).
159) Phillips, G. M., Hunter, J. S., Sutton, L. E.: J. Chem. Soc. *1945*, 146.
160) Brumberger, H., Marcus, R. A.: J. Chem. Phys. *24*, 741 (1956).
161) Drake, J. E., Simpson, J.: Inorg. Nucl. Chem. Letters *3*, 87 (1967).
162) Fritz, G.: Z. Naturforsch. *8b*, 776 (1953).
163) Fritz, G.: Z. Anorg. Allgem. Chem. *280*, 332 (1955).
164) Gokhale, S. D., Jolly, W. L.: Inorg. Chem. *4*, 596 (1965).

165) Gokhale, S. D., Jolly, W. L.: Inorg. Chem. *3*, 1141 (1964).
166) Aylett, B. J., Emeléus, H. J., Maddock, A. G.: J. Inorg. Nucl. Chem. *1*, 187 (1955).
167) Amberger, E., Boeters, H. D.: Chem. Ber. *97*, 1999 (1964).
168) Beagley, B., Robiette, A. G., Sheldrick, G. M.: Chem. Commun. *1967*, 601.
169) Beagley, B.: Chem. Commun. *1966*, 388.
170) Finholt, A. E., Helling, C., Imhof, V., Nielsen, L., Jacobson, E.: Inorg. Chem. *2*, 504 (1963).
171) Messinger, J., Engels, C.: Ber. Deut. Chem. Ges. *21*, 326 (1888).
172) Messinger, J., Engels, C.: Ber. Deut. Chem. Ges. *21*, 2919 (1888).
173) de Girard, J.: Ann. Chim. Phys., Ser. VII, *2*, 2 (1884).
174) Hoffman, A.: J. Am. Chem. Soc. *43*, 1684 (1921).
175) Hoffman, A.: J. Am. Chem. Soc. *52*, 2995 (1930).
176) Reeves, W. A., Flynn, F. F., Guthrie, J. D.: J. Am. Chem. Soc. *77*, 3923 (1955).
177) Horak, J., Ettel, V.: Collection Czech. Chem. Commun. *26*, 2401 (1961).
178) Raver, K. R., Bruker, A. B., Soborovskii, L. Z.: Zh. Obshch. Khim. *32*, 588 (1962); C.A. *58*, 6857 (1963).
179) Raver, K. R., Soborovskii, L. Z.: SSSR Pat., 143 395; C.A. *57*, 9882 (1962).
180) Bruker, A. B., Baranaev, M. K., Grinshtein, E. I., Novoselova, R. I., Prokhorova, V. V., Soborovskii, L. Z.: Zh. Obshch. Khim. *33*, 1919 (1963); C.A. *59*, 11207 (1963).
181) Emeléus, H. J.: J. Chem. Soc. *1954*, 2979.
182) Coates, H., Hoye, P. A.: Brit. Pat., 854.182 (1960).
183) Buckler, S. A., Doll, L.: U.S. Pat., 2.999.882 (1959).
184) de Girard, A.: Ann. Chim. [6] *2*, 11 (1884).
185) Buckler, S. A.: U.S. Pat., 3.013.085 (1961); C.A. *57*, 11236 (1962).
186) Buckler, S. A., Wystrach, V. P.: J. Am. Chem. Soc. *80*, 6454 (1958).
187) Buckler, S. A., Wystrach, V. P.: J. Am. Chem. Soc. *83*, 168 (1961).
188) Hart, F. A., Mann, F. G.: J. Chem. Soc. *1955*, 4107.
189) Buckler, S. A.: J. Am. Chem. Soc. *82*, 4215 (1960).
190) Ettel, V., Horak, J.: Collection Czech. Chem. Commun. *26*, 1949 (1961).
191) Buckler, S. A., Epstein, M.: J. Am. Chem. Soc. *82*, 2076 (1960).
192) Buckler, S. A., Epstein, M.: Tetrahedron *18*, 1211 (1962).
193) Buckler, S. A., Epstein, M.: Tetrahedron *18*, 1221 (1962).
194) Parshall, G. W.: Inorg. Chem. *4*, 52 (1965).
195) Epstein, M., Buckler, S. A.: J. Am. Chem. Soc. *83*, 3279 (1961).
196) Stiles, A. R., Rust, F. F., Vaughan, W. E.: J. Am. Chem. Soc. *74*, 3282 (1952); U.S. Pat. 2.803.597 (1959).
197) Rauhut, M. M., Currier, H. A., Semsel, A. M., Wystrach, V. P.: J. Org. Chem. *26*, 5138 (1961).
198) Bereslavsky, E. V.: U.S. Pat. 2.797.153 (1957).
199) Brown, H. C.: U.S. Pat. 2.584.112 (1952).
200) Rauhut, M. M., Hechenbleikner, I., Currier, H. A., Schaefer, F. C., Wystrach, V. P.: J. Am. Chem. Soc. *81*, 1103 (1959).
201) Hoff, M. C., Hill, P.: J. Org. Chem. *24*, 356 (1959).
202) Ingold, C. K.: Structure and mechanism in organic chemistry, p. 691. Ithaca, N. Y.: Cornell University Press 1953.
203) Burch, G. M., Goldwhite, H., Haszeldine, R. N.: J. Chem. Soc. *1963*, 1083.
204) Fields, R., Goldwhite, H., Haszeldine, R. N., Kirman, J.: J. Chem. Soc. *1966*, 2075.
205) Parshall, G. W., England, D. C., Linsey, R. V.: J. Am. Chem. Soc. *81*, 4801 (1959).
206) England, D. C., Parshall, G. W.: U.S. Pat. 2.879.302 (1959).
207) Buckler, S. A.: J. Org. Chem. *24*, 1460 (1959).
208) Buckler, S. A.: U.S. Pat. 2.969.390 (1961).

209) Papp, G. P., Buckler, S. A.: J. Org. Chem. *31*, 588 (1966).
210) Hunter, R. F.: Chem. News *1930*, 50.
211) Hieber, W., Winter, E.: Chem. Ber. *97*, 1037 (1964).
212) Coates, G. E., Whitcombe, R. A.: J. Chem. Soc. *1956*, 3351.
213) Didchenko, R., Alix, J. E., Toeniskoetter, R. M.: J. Inorg. Nucl. Chem *14*, 35 (1960).
214) Job, A., Dusollier, G.: Compt. Rend. *184*, 1454 (1927).
215) Lecoq, H.: Bull. Soc. Chim. Belges *42*, 199 (1933).
216) Mann, F. G., Miller, I. T.: J. Chem. Soc. *1952*, 3039.
217) Tyka, R., Plazek, E.: Bull. Acad. Polon. Sci., Ser. Sci. Chim. *9*, 577 (1961).
218) Plazek, E., Tyka, R.: Roczniki Chem. *33*, 549 (1959).
219) Lawless, J. J., Searle, H. T.: J. Chem. Soc. *1962*, 4200.
220) Svehla, P.: Collection Czech. Chem. Commun. *31*, 4712 (1966); C.A. *66*, 61329 (1967).
221) Paris, R., Tardy, P.: Compt. Rend. *223*, 242 (1946).
222) Horak, J., Ettel, V.: Sb. Vysoke Skoly Chem.-Technol. Praze, Org. Technol *5*, 93 (1960); C.A. *62*, 12440 (1965).
223) Haszeldine, R. N., Kidd, J. M.: J. Chem. Soc. *1953*, 3219.
224) Emeléus, H. J., Nabi, S. N.: J. Chem. Soc. *1960*, 1103.
225) Flury, F., Zernik, F.: Schädliche Gase, S. 170. Berlin 1931.
226) Flury, F.: Arch. Exptl. Pathol. Pharmakol. *138*, 71 (1928).
227) Kloos, E. J., Spinetti, L., Raymond, L. D.: U.S. Bureau of Mines, Inform. Circ. Nr. 8291, 7 (1966).
228) Wiesner, H.: Monatsschr. Kinderheilk. *105*, 312 (1957).
229) Hallermann, W., Pribilla, O.: Arch. Toxikol. *17*, 219 (1959).
230) Klimmer, O. R.: Arch. Toxikol. *24*, 164 (1969).
231) Trautz, M., Bhandarkar, D. S.: Z. Anorg. Allgem. Chem. *106*, 95 (1919).
232) Hinshelwood, C. N., Topley, B.: J. Chem. Soc. *125*, 393 (1924).
233) Melville, H. W., Roxburgh, H. L.: J. Chem. Soc. *1933*, 586.
234) Trautz, M., Gabler, W.: Z. Anorg. Allgem. Chem. *180*, 321 (1929).
235) Shantarovich, P. S.: Acta Physiochim. USSR *6*, 65 (1937); C.A. *31*, 7304 (1937).
236) Melville, H. W., Roxburgh, H. L.: J. Chem. Soc. *1934*, 264.
237) Andreev, E. A., Kavtaradze, N. N.: Probl. Kinetiki i Kataliza, Akad. Nauk SSSR *6*, 293 (1949); C.A. *47*, 5776 (1953).
238) Andreev, E. A., Kavtaradze, N. N.: Izv. Akad. Nauk SSSR, Otd. Khim. Nauk *1952*, 1021; C.A. *47*, 5777 (1953).
239) Dalton, R. H., Hinshelwood, C. N.: Proc. Roy. Soc. (London) A *125*, 294 (1929).
240) Dalton, R. H.: Proc. Roy. Soc. (London) A *128*, 263 (1930).
241) Gray, S. C., Melville, H. W.: Trans. Faraday Soc. *31*, 452 (1935).
242) Melville, H. W., Roxburgh, H. L.: J. Chem. Phys. *2*, 739 (1934).
243) Bushmakin, I. N., Vvedenskii, A. A., Frost, A. V.: J. Gen. Chem. (USSR) *2*, 415 (1932); C.A. *27*, 1806 (1933).
244) Bushmakin, I. N., Frost, A. V.: J. Appl. Chem. (USSR) *6*, 607 (1933); C.A. *28*, 3677 (1934).
245) Bendall, J. R., Mann, F. G., Purdie, D.: J. Chem. Soc. *1942*, 157.
246) Wichelhaus, H.: Ber. Deut. Chem. Ges. *38*, 1725 (1905).
247) Weyl, T.: Ber. Deut. Chem. Ges. *39*, 4340 (1906).
248) Buckler, S. A., Doll, L., Lind, F. K., Epstein, M.: J. Org. Chem. *27*, 794 (1962).
249) Guenebaut, H., Pascat, B.: J. Chim. Phys. *61*, 592 (1964).
250) Guenebaut, H., Pascat, B.: Compt. Rend. *267*, 677 (1963).
251) Guenebaut, H., Pascat, B.: Compt. Rend. *295*, 2412 (1964).
252) Guenebaut, H., Pascat, B., Berthou, J. M.: J. Chim. Phys. *62*, 867 (1965).

253) Wiles, D. M., Winkler, C. A.: J. Phys. Chem. *61*, 620 (1957).
254) Guenebaut, H., Pascat, B.: Compt. Rend. *256*, 2850 (1963).
255) Norrish, R. G. W., Oldershaw, G. A.: Proc. Roy. Soc. (London) *A 262*, 10 (1961).
256) Moser, L., Brukl, A.: Z. Anorg. Allgem. Chem. *121*, 78 (1922).
257) Brukl, A.: Z. Anorg. Allgem. Chem. *125*, 252 (1922).
258) Scholder, R., Apel, A., Haken, H. L.: Z. Anorg. Allgem. Chem. *232*, 1 (1937).
259) Birchall, T., Jolly, W. L.: Inorg. Chem. *5*, 2177 (1966).
260) Rudolph, R. W., Parry, R. W.: J. Am. Chem. Soc. *89*, 1621 (1967).
261) Borde, C., Henry, A., Henry, L.: Compt. Rend. Acad. Sci. Paris, Ser. A, B. 263 B, 619 (1966).
262) Halmann, M., Platzner, I.: J. Phys. Chem. *71*, 4522 (1967).
263) Drake, J. E., Simpson, J.: J. Chem. Soc. (London) *A 1968*, 974.
264) Fourcroy, A. F., de Vauquelin, L. D.: Ann. Chim. (Paris) *21*, 189 (1797).
265) Davy, H.: Phil. Trans. Roy. Soc. (London) *1809*, 39, 450.
266) Schneider, W. G., Bernstein, H. J., Pople, J. A.: J. Chem. Phys. *28*, 601 (1958).
267) Staveley, L. A. K., Tupman, W. J.: J. Chem. Soc. *1950*, 3597.
268) Durrant, A. A., Pearson, T. G., Robinson, P..L.: J. Chem. Soc. *1934*, 730.
269) Briner, E.: J. Chim. Phys. *4*, 476 (1906).
270) Berl, E.: Chem. Met. Eng. *53*, 130 (1946).
271) Kordes, E.: Z. Elektrochem. *57*, 731 (1953).
272) Pickering, S. F.: Natl. Bur. Std. (U.S.) Circ. No. 279,1 (1926).
273) Leduc, A., Sacerdote, P.: Compt. Rend. *125*, 379 (1897).
274) Fritz, G.: Z. Naturforsch. *8b*, 776 (1953).
275) Fritz, G.: Z. Anorg. Allgem. Chem. *280*, 332 (1955).
276) Leverrier, U. J. J.: Ann. Chim. Phys. *60*, 174 (1835).
277) Amato, D.: Gazz. Chim. Ital. *14*, 57 (1884).
278) Retgers, J. W.: Z. Anorg. Allgem. Chem. *7*, 265 (1894).
279) Inorganic syntheses, Vol. II, p. 141. New York: McGraw-Hill Book Co. 1946.
280) Legoux, C.: Compt. Rend. *209*, 47 (1939).
281) de Guye, P.: Bull. Soc. Chim. France *5*, 339 (1909).
282) Matheson, G. L., Maass, O.: J. Am. Chem. Soc. *51*, 674 (1929).
283) Rose, H.: Ann. Physik [2] *24*, 109 (1832).
284) Kelley, K. K.: U.S. Bur. Mines Bull. Nr. 383, 1 (1935).
285) Dalton, J.: Ann. Phil. Thomson [2] *11*, 7 (1818).
286) Steacie, E. W. R., McDonald, R. D.: Can. J. Res. *12*, 711 (1924).
287) Horner, L., Beck, P., Hoffmann, H.: Chem. Ber. *92*, 2088 (1959).
288) Aylett, B. J., Emeléus, H. J., Maddock, A. G.: J. Inorg. Nucl. Chem. *1*, 187 (1955).
289) Burg, A. B., Wagner, I.: J. Am. Chem. Soc. *75*, 3872 (1953).
290) Martin, D. R., Dial, R. E.: J. Am. Chem. Soc. *72*, 852 (1950).
291) Wiberg, E., Heubaum, U.: Z. Anorg. Allgem. Chem. *225*, 270 (1935).
292) Parry, R. W., Bissot, T. C.: J. Am. Chem. Soc. *78*, 1524 (1956).
293) Besson, A.: Compt. Rend. *110*, 80 (1890).
294) Besson, A.: Compt. Rend. *110*, 516 (1890).
295) Waddington, T. C., Klanberg, F.: J. Chem. Soc. *1960*, 2332.
296) Wartik, T., Apple, E. F.: J. Am. Chem. Soc. *80*, 6155 (1958).
297) Besson, A.: Compt. Rend. *113*, 78 (1891).
298) Fischer, A.: Z. Naturforsch. *13a*, 105 (1958).
299) Höltje, R.: Z. Anorg. Allgem. Chem. *190*, 241 (1930).
300) Besson, A.: Compt. Rend. *110*, 240 (1890).
301) Stock, A., Böttcher, W., Lenger, W.: Ber. Deut. Chem. Ges. *42*, 2839 (1909).
302) Waddington, T. C., Nabi, S. N.: Proc. Pakistan Sci. Conf. *12*, Pt 13, C 7 (1960);

C.A. *56*, 9678 (1962).
303) Royen, P., Hill, K.: Z. Anorg. Allgem. Chem. *229*, 112 (1936).
304) Besson, A.: Compt. Rend. *110*, 1258 (1890).
305) Janovsky, J. V.: Ber. Deut. Chem. Ges. *8*, 1936 (1875).
306) Gutmann, V.: Z. Anorg. Allgem. Chem. *266*, 331 (1951).
307) Ogier, J.: Bull. Soc. Chim. France [2] *32*, 483 (1879); Compt. Rend. *89*, 705 (1879); Ann. Chim. Phys. [5] *20*, 5 (1880).
308) Skinner, S.: Proc. Roy. Soc. (London) *42*, 283 (1887).
309) Sérullas, G. S.: Ann. Chim. Phys. *48*, 87 (1831).
310) Reuter, M., Orthner, L.: DBP 1041957 (1958); C.A. *55*, 1444 (1961).
311) de Fourcroy, A. F.: Ann. Chim. Phys. *4*, 249 (1790).
312) Horak, J.: Chem. Listy *55*, 1278 (1961).
313) Stock, A.: Ber. Deut. Chem. Ges. *53*, 837 (1920).
314) Solovev, V. K.: Gorn. Zh. *115*, 34 (1939); Chem. Zentralbl. *1940*, I, 3967.
315) Devyatykh, G. G., Ezheleva, A. E., Zorin, A. D., Zueva, M. V.: Zh. Neorgan. Khim. *8*, 1307 (1963).
316) Hofmann, A. W.: Liebigs Ann. Chem. *103*, 355 (1857).
317) Holt, A., Myers, J. E.: Z. Anorg. Allgem. Chem. *82*, 278 (1913).
318) Corbridge, D. E. C.: Topics in phosphorus chemistry, Vol. 3, p. 91. New York: Interscience Publ. 1966.
319) Durrant, A. A., Pearson, T. G., Robinson, P. L.: J. Chem. Soc. *1934*, 730.
320) Steele, B. D., McIntosh, D.: Z. Phys. Chem. *55*, 140 (1906).
321) Dobinski, S.: Z. Physik *83*, 129 (1933).
322) Cauquil, G.: J. Chim. Phys. *24*, 53 (1927).
323) Reuter, M., Orthner, L.: DBP 1035135 (1958); C.A. *54*, 14125 (1960).
324) Coniglio, L., Caglioti, V.: Rend. Accad. Sci. Fis. Mat. Soc. Nazl. Sci. Napoli *33*, 154 (1927).
325) Dickinson, R. G.: J. Am. Chem. Soc. *44*, 1489 (1922).
326) Levy, H. A., Peterson, S. W.: J. Am. Chem. Soc. *75*, 1536 (1953).
327) Gopal, N.: Indian J. Phys. *7*, 285 (1932).
328) Heinemann, A.: Ber. Bunsenges. Physik. Chem. *68*, 280 (1964).
329) Zugravescu, P. G., Zugravescu, M. A.: Rev. Chim. (Bucharest) *17*, 704 (1966); C.A. *66*, 101326 (1967).
330) Taylor, R. W. D.: Chem. Ind. (London) *33*, 1116 (1968).
331) Kobayashi, Y., Meguro, T.: Bunseki Kagaku *16*, 1359 (1967); C.A. *68*, 107679 (1968).
332) Leffler, A. J., Teach, E. G.: J. Am. Chem. Soc. *82*, 2710 (1960).
333) Ettel, V., Horak, J.: Collection Czech. Chem. Commun. *26*, 2087 (1961).
334) Wartik, T., Apple, E. F.: J. Am. Chem. Soc. *80*, 6155 (1958).
335) Fischer, E. O., Louis, E., Schneider, R. J. J.: Angew. Chem. *80*, 122 (1968).
336) Klanberg, F., Muetterties, E. L.: J. Am. Chem. Soc. *90*, 3296 (1968).
337) Campbell, J. M., Stone, F. G. A.: Angew. Chem. *81*, 120 (1969).
338) Boyd, D. B., Lipscomb, W. N.: J. Chem. Phys. *46*, 910 (1967).
339) Siebert, H., Eints, J., Fluck, E.: Z. Naturforsch. *23b*, 1006 (1968).
340) Fischer, E. O., Louis, E., Kreiter, C. G.: Angew. Chem. *81*, 397 (1969).
341) Devyatykh, G. G., Zorin, A. D., Postnikova, T. K., Umilin, V. A.: Zh. Neorgan. Khim. *14*, 1626 (1969); C.A. *71*, 64783 (1969).
342) Zorin, A. D., Runovskaya, I. V., Lyakhmanov, S. B., Yudanova, L. V.: Zh. Neorgan. Khim. *12*, 2529 (1967); C.A. *68*, 43402 (1968).
343) Vlasov, S. M., Devyatykh, G. G.: Zh. Neorgan. Khim. *11*, 2681 (1966); C.A. *67*, 5929 (1967).

344) Rankine, A. O., Smith, C. J.: Phil. Mag. *42*, 601 (1921).
345) Helminger, P., Gordy, W.: Phys. Rev. *188*, 100 (1969).
346) Lehn, J. M., Munsch, B.: J. Chem. Soc. (London) *D 1969*, 1327.
347) Hillier, I. H., Saunders, V. R.: J. Chem. Soc. (London) *D 1970*, 316.
348) Raevskii, O. A., Khalitov, F. G.: Izv. Akad. Nauk SSSR, Ser. Khim. *1970*, 2368; C.A. *74*, 117213 (1971).
349) Devyatykh, G. G., Zorin, A. D., Runovskaya, I. V.: Dokl. Akad. Nauk SSSR *188*, 1082 (1969); C.A. *72*, 47751 (1970).
350) Armstrong, R. L., Courtney, J. A.: J. Chem. Phys. *51*, 457 (1969).
351) Frost, D. C., McDowell, C. A., Sandhu, J. S., Vroom, D. A.: Advan. Mass Spectrometry *1968*, 781.
352) Branton, G. R., Frost, D. C., McDowell, C. A., Stenhouse, I. A.: Chem. Phys. Letters *5*, 1 (1970).
353) Giardini-Guidoni, A., Volpi, U. G. G.: Nuovo Cimento *17*, 919 (1960).
354) Eyler, J. R.: Inorg. Chem. *9*, 981 (1970).
355) Basco, N., Yee, K. K.: Nature *216*, 998 (1967); C.A. *68*, 25388 (1968).
356) Olah, G. A., McFarland, C. W.: J. Org. Chem. *34*, 1832 (1969).
357) Graham, T.: Phil. Mag. 3 *5*, 401 (1834).
358) Fujioka, G. S., Cady, G. H.: J. Am. Chem. Soc. *79*, 2451 (1957).
359) Holtz, D., Beauchamp, J. L.: J. Am. Chem. Soc. *91*, 5913 (1969).
360) Beauchamp, J. L., Buttrill, S. E.: J. Chem. Phys. *48*, 1783 (1968).
361) Haney, M. A., Franklin, J. L.: J. Phys. Chem. *73*, 4328 (1969).
362) Haney, M. A., Franklin, J. L.: J. Chem. Phys. *50*, 2028 (1969).
363) Guyon, P.-M.: J. Chim. Phys. *66*, 468 (1969).
364) Berechnet aus Daten von Harrison, A. G., Ivko, A., Van Raalte, D.: Can. J. Chem. *44*, 1625 (1966) und Refaey, K. M. A., Chupka, W. A.: J. Chem. Phys. *48*, 5205 (1968).
365) Ionization potentials, appearance potentials, and heats of formation of gaseous positive ions, NSRDS–NBS 26. Washington, D.C.: U.S. Government Printing Office 1969.
366) Vedeneev, V. I., Gurvich, L. V., Kondratev, V. N., Medvedev, V. A., Frankevich, Ye. L.: Bond energies, ionization potentials, and electron affinities. New York, N.Y.: Scripta Technica Ltd., St. Martin's Press 1966.
367) Hild, K., Heidemann, W.: Beckman Rep. 3/4,12 (1966); C.A. *70*, 74030 (1969).
368) Dumas, T.: J. Agr. Food Chem. *17*, 1164 (1969); C.A. *71*, 128321 (1969).
369) Berck, B., Westlake, W. E., Gunther, F. A.: J. Agr. Food Chem. *18*, 143 (1970).
370) Agranov, Kh. I.: Nov. Obl. Prom., Sanit. Khim. *1969*, 60; C.A. *71*, 116275 (1969).
371) Sonobe, K., Nakaoka, T.: Japan. Pat. 26387 (1968); C.A. *71*, 5007 (1969).
372) Devyatykh, G. G., Kedyarkin, V. M., Zorin, A. D.: Zh. Neorgan. Khim *14*, 2011 (1969); C.A. *71*, 97898 (1969).
373) Trautz, M., Bhandarkar, H.: Z. Anorg. Allgem. Chem. *106*, 95 (1919).
374) Hinshelwood, C. N., Topley, B.: J. Chem. Soc. *125*, 393 (1924).
375) van't Hoff, J. H., Kooj, D. M.: Z. Phys. Chem. *12*, 125 (1893).
376) Strater, K., Mayer, A.: Semicond. Silicon, Internat. Sympos. Pap. 1st, *1969*, 469; C.A. *71*, 108630 (1969).
377) Davies, P. B., Trush, B. A.: Proc. Roy. Soc. Ser. A *302*, 243 (1967).
378) Heckmann, G., Fluck, E.: Z. Naturforsch. *24b*, 953 (1969).
379) Heckmann, G., Fluck, E.: Z. Naturforsch. *24b*, 1092 (1969).
380) Heckmann, G., Fluck, E.: Z. Naturforsch. *25b*, 1226 (1970).
381) Heckmann, G., Fluck, E.: Z. Naturforsch. *26b*, 63 (1971).
382) Heckmann, G., Fluck, E.: Z. Naturforsch. *26b*, 282 (1971).
383) Fluck, E.: Die kernmagnetische Resonanz und ihre Anwendung in der anorganischen Chemie. Berlin-Heidelberg-New York: Springer 1963.

61

384) Heckmann, G., Fluck, E.: Mol. Phys. *23*, 175 (1972).
385) Gutowsky, H. S., McCall, D. W., Slichter, C. P.: J. Chem. Phys. *21*, 279 (1953).
386) Gutowsky, H. S., McCall, D. W.: J. Chem. Phys. *22*, 162 (1954).
387) Van Wazer, J. R., Callis, C. F., Shoolery, J. N., Jones, R. C.: J. Am. Chem. Soc. *78*, 5715 (1956).
388) Parks, J. R.: J. Am. Chem. Soc. *79*, 757 (1957).
389) Morin, C.: Bull. Soc. Chim. (France) *1961*, 1446.
390) Jones, R. A. Y., Katritzky, A. R.: Angew. Chem. *74*, 60 (1962).
391) Manatt, S. L., Juvinall, G. L., Elleman, D. D.: J. Am. Chem. Soc. *85*, 2664 (1963).
392) Schumann, H., Stelzer, O., Kuhlmey, J., Niederreuther, U.: J. Organometal. Chem. *28*, 105 (1971).
393) Fluck, E.: Chemiker-Ztg. *94*, 833 (1970).
394) Fluck, E., Bürger, H., Götze, U.: Z. Naturforsch. *22b*, 912 (1967).
395) Engelhardt, G., Reich, P., Schumann, H.: Z. Naturforsch. *22b*, 352 (1967).
396) Baudler, M., Ständeke, H., Dobbers, J., Borgardt, M., Strabel, H.: Naturwissenschaften *53*, 251 (1966).
397) Wiberg, E., Müller-Schiedmayer, G.: Z. Anorg. Allgem. Chem. *308*, 352 (1961).
398) Ramirez, F., Aguiar, A.: 134. Meeting Am. Chem. Soc., 42N, Sept. 1958.
399) Rabinowitz, J.: Helv. Chim. Acta *53*, 53 (1970).
400) Rabinowitz, J., Woeller, F., Flores, J., Krebsbach, R.: Nature (London) *224*, 796 (1969).
401) Buchanan, J. W., Hanrahan, R. J.: Radiation Res. *44*, 296 (1970); C.A. *74*, 17986 (1971).
402) Buchanan, J. W., Hanrahan, R. J.: Radiation Res. *42*, 244 (1970); C.A. *72*, 138289 (1970).
403) Sisler, H. H., Sarkis, A., Ahujo, H. S., Drago, R. J., Smith, N. L.: J. Am. Chem. Soc. *81*, 2982 (1959).
404) Hart, W. A., Sisler, H. H.: Inorg. Chem. *3*, 617 (1964).
405) Clemens, D. F., Sisler, H. H.: Inorg. Chem. *4*, 1222 (1965).
406) Vetter, H. J., Nöth, H.: Z. Anorg. Allgem. Chem. *330*, 233 (1964).
407) Jain, S. R., Krannich, L. K., Highsmith, R. E., Sisler, H. H.: Inorg. Chem. *6*, 1058 (1967).
408) Highsmith, R. E., Sisler, H. H.: Inorg. Chem. *7*, 1740 (1968).
409) Petrov, K. A., Parshina, V. A., Orlov, B. A., Tsypina, G. M.: Zh. Obshch. Khim. *32*, 4017 (1962); J. Gen. Chem. USSR *32*, 3944 (1962).
410) Jaura, K. L., Maini, B. K., Kaushik, R. L.: Res. Bull. Panjab Univ. *18*, 165 (1967); C.A. *69*, 26738 (1968).
411) Sawodny, W., Goubeau, J.: Z. Anorg. Allgem. Chem. *356*, 289 (1968).
412) Davis, J., Drake, J. E.: J. Chem. Soc. A *1970*, 2959.
413) Manatt, S. L., Junivall, G. L., Ellemann, D. D.: J. Am. Chem. Soc. *85*, 2664 (1963).
414) Sheldrick, G. M.: Trans. Faraday Soc. *63*, 1077 (1967).
415) Glidewell, C., Sheldrick, G. M.: J. Chem. Soc. A *1969*, 350.
416) Ebsworth, E. A. V., Glidewell, C., Sheldrick, G. M.: J. Chem. Soc. A *1969*, 352.
417) Davis, J., Drake, J. E., Goddard, N.: J. Chem. Soc. A *1970*, 2962.
418) Strizhevskii, I. I., Slizovskaya, L. V.: Svarochn. Proizvod. *1968*, 43; C.A. *70*, 79530 (1969).
419) Whistler, R. L., Wang, Chih-Cheng, Inokawa, S.: J. Org. Chem. *33*, 2495 (1968).
420) Evers, C., Street, E. H.: J. Am. Chem. Soc. *78*, 5726 (1956).
421) Royen, P., Hill, K.: Z. Anorg. Allgem. Chem. *229*, 97 (1936).
422) Gattermann, L., Haussknecht, W.: Ber. Deut. Chem. Ges. *23*, 1174 (1890).
423) Gunn, S. R., Green, L. G.: J. Phys. Chem. *65*, 779 (1961).
424) Nixon, E. R.: J. Phys. Chem. *60*, 1054 (1956).

[425] Baudler, M., Schmidt, L.: Z. Anorg. Allgem. Chem. *289*, 219 (1957).
[426] Wheatley, P. J.: J. Chem. Soc. *1956*, 4514.
[427] Lynden-Bell, R. M.: Trans. Faraday Soc. *57*, 888 (1961).
[428] Thénard, P.: Compt. Rend. *18*, 652 (1844).
[429] Hofmann, A. W.: Ber. Deut. Chem. Ges. *7*, 530 (1874).
[430] Retgers, J. W.: Naturw. Rundschau *10*, 384 (1895).
[431] Datta, J.: J. Indian Chem. Soc. *29*, 751 (1952).
[432] Wiles, D. M., Winkler, C. A.: J. Phys. Chem. *61*, 620 (1957).
[433] Michaelis, A., Pitsch, M.: Liebigs Ann. Chem. *310*, 45 (1900).
[434] Beichl, G. J., Evers, E. C.: J. Am. Chem. Soc. *80*, 5344 (1958).
[435] Berthelot, D., Gaudechon, H.: Compt. Rend. *156*, 1243 (1913).
[436] Baudler, M., Ständeke, H., Borgardt, M., Strabel, H., Dobbers, J.: Naturwissenschaften *53*, 106 (1966).
[437] Baudler, M., Schmidt, L.: Naturwissenschaften *46*, 577 (1959).
[438] Royen, P., Rocktäschel, C., Mosch, W.: Angew. Chem. *76*, 860 (1964).
[439] Gmelins Handbuch der anorganischen Chemie, 8th edit., System No. 16, Part C, p. 53. Weinheim: Verlag Chemie 1965.
[440] Baudler, M., Ständeke, H., Dobbers, J.: Z. Anorg. Allgem. Chem. *353*, 122 (1967).
[441] White, W. E., Bushey, A. H.: J. Am. Chem. Soc. *66*, 1666 (1944).
[442] Welker, H.: Z. Naturforsch. *7a*, 744 (1952).
[443] Addamiano, A.: Acta Cryst. *13*, 505 (1960).
[444] Stackelberg, v. M., Paulus, R.: Z. Physik.Chem. (B) *22*, 305 (1933).
[445] Zintl, E., Husemann, E.: Z. Physik.Chem. (B) *21*, 138 (1933).
[446] Stackelberg, v. M.: Z. Physik. Chem. (B) *27*, 53 (1934).
[447] Engelhardt, G.: Z. Anorg. Allgem. Chem. *387*, 52 (1972).
[448] Cradock, S., Ebsworth, E. A. V., Davidson, G., Woodward, L. A.: J. Chem. Soc. (London) *A 1967*, 1229.
[449] Marsmann, H., Groenweghe, L. C. D., Schaad, L. J., Van Wazer, J. R.: J. Am. Chem. Soc. *92*, 6107 (1970).
[450] Mitchell, K. A. R.: Can. J. Chem. *46*, 3499 (1968).
[451] Fischler, J., Halmann, M.: J. Chem. Soc. *1964*, 31.
[452] Halmann, M., Kugel, L.: J. Inorg. Nucl. Chem. *25*, 1343 (1963).
[453] Halmann, M.: Chem. Rev. *64*, 689 (1964).
[454] Halmann, M., Kugel, L.: J. Chem. Soc. *1964*, 4025.
[455] Halmann, M.: Chemical effects of nuclear transformations, p. 195. Vienna: Int. Atomic Energy Agency 1961.
[456] Mitteilung VII der Kommission zur Prüfung gesundheitsschädlicher Arbeitsstoffe, Deutsche Forschungsgemeinschaft 1971.
[457] Rose, H.: Ann. Physik [2] 6, 199 (1826); *24*, 295 (1832).
[458] Rubenovitch, E.: Compt. Rend. *128*, 1398 (1899).
[459] Riban, J.: Compt. Rend. *88*, 581 (1879); Bull. Soc. Chim. France [2] *31*, 385 (1897).
[460] Scholder, R., Pattock, K.: Z. Anorg. Allgem. Chem. *220*, 250 (1934).
[461] Höltje, R., Schlegel, H.: Z. Anorg. Allgem. Chem. *243*, 246 (1940).
[462] Moser, L., Brukl, A.: Z. Anorg. Allgem. Chem. *121*, 73 (1922).
[463] Boettger, R.: Beitr. Phys. Chem. *2*, 116 (1840).
[464] Schönberg, N.: Acta Chem. Scand. *8*, 226 (1954).
[465] Guenebaut, H., Pascat, B.: Compt. Rend. *255*, 1741 (1962).
[466] Dulong, P. L.: Ann. Chim. Phys. [2] *31*, 154 (1826).
[467] Bleekrode, L.: Proc. Roy. Soc. (London) *37*, 339 (1884).
[468] Smith, C.: Proc. Roy. Soc. (London) *A 87*, 366 (1912).
[469] Barter, C., Meisenheimer, R. G., Stevenson, D. P.: J. Phys. Chem. *64*, 1312 (1960).

470) Mallemann de, R., Gabiano, P.: Compt. Rend. *199*, 600 (1934).
471) Weston, R. E.: J. Am. Chem. Soc. *76*, 1027 (1954).
472) Cailletet, L., Bordet, L.: Compt. Rend. *95*, 58 (1882).
473) Skinner, S.: Proc. Roy. Soc. (London) *42*, 283 (1887).
474) Claussen, W. F.: J. Chem. Phys. *19*, 1425 (1951).
475) Stackelberg, v. M., Müller, H. R.: Z. Elektrochem. *58*, 25 (1954).
476) Powell, H. M.: J. Chem. Soc. *1954*, 2658.
477) Melville, H. W.: Proc. Roy. Soc. (London) A *139*, 541 (1933).
478) Melville, H. W., Bolland, J. L., Roxburgh, H. L.: Proc. Roy. Soc. (London) A *160*, 406 (1937).
479) Hinshelwood, C. N., Clusius, K.: Proc. Roy. Soc. (London) A *129*, 589 (1930).
480) Fluck, E.: In press.
481) Drummond, D. H.: J. Am. Chem. Soc. *49*, 1901 (1927).
482) Ipatiev, W. N., Frost, A. W.: Ber. Deut. Chem. Ges. *63*, 1104 (1930); Zh. Russ. Fiz.-Khim. Obshchestva Chast'Khim. *62*, 1123 (1930).
483) Sanderson, R. T.: J. Chem. Phys. *20*, 535 (1952).
484) Kooij, D. M.: Z. Physik. Chem. *12*, 155 (1893).
485) Dushmann, S.: J. Am. Chem. Soc. *43*, 397 (1921).
486) Yamazaki, E.: J. Tokyo Chem. Soc. *40*, 606 (1919); C.A. *1919*, 3053.
487) Bodenstein, M.: Z. Elektrochem. *35*, 535 (1929).
488) Barber, R. M.: Trans. Faraday Soc. *32*, 490 (1936).
489) Laidler, K. J., Glasstone, S., Eyring, H.: J. Chem. Phys. *8*, 667 (1940).
490) Lewis, W. C. M.: Phil. Mag. [6] *39*, 26 (1920).
491) Roy, S. C.: Proc. Roy. Soc. (London) A *110*, 543 (1926).
492) Temkin, M.: Acta Physicochim. USSR *8*, 141 (1938).
493) Kharasch, M. S., Reinmuth, O.: Grignard reactions of molecular substances, p. 1335. New York: Prentice Hall 1954.

Received February 29, 1972

Transition Metal Dithio- and Diselenophosphate Complexes

Prof. Dr. John R. Wasson, Dr. Gerald M. Woltermann and Dr. Henry J. Stoklosa

Department of Chemistry, University of Kentucky, Lexington, Kentucky, USA

Contents

I. Introduction

Within recent years the chemistry of compounds containing metal-sulfur bonds has attracted increasing attention. This is partly due to the unusual properties of many of the complexes with sulfur-containing ligands which pose a challenge to interpretation. In part, the interest has also been due to the relevancy of the compounds to problems in biochemistry and, particularly for the compounds discussed here, uses of the complexes as antioxidants, oil additives, coloring agents for plastics, and as pesticides. Several reviews of metal complexes of ligands containing sulfur and selenium donor atoms have appeared [1-19] but none of these has been exclusively concerned with metal dithio- and diselenophosphate complexes.

Fig. 1 provides a listing of the various sulfur donor ligands whose complexes have been the subject of considerable research. The list of ligands in Fig. 1 is not exhaustive since only potentially bidentate ligands are given. The electronic properties of complexes with bidentate sulfur donor ligands are usually similar although their physical properties, e.g., solubility, can vary widely. This review deals primarily with dithio- and diselenophosphate complexes; however, related complexes are discussed wherever they are pertinent to the present discussion.

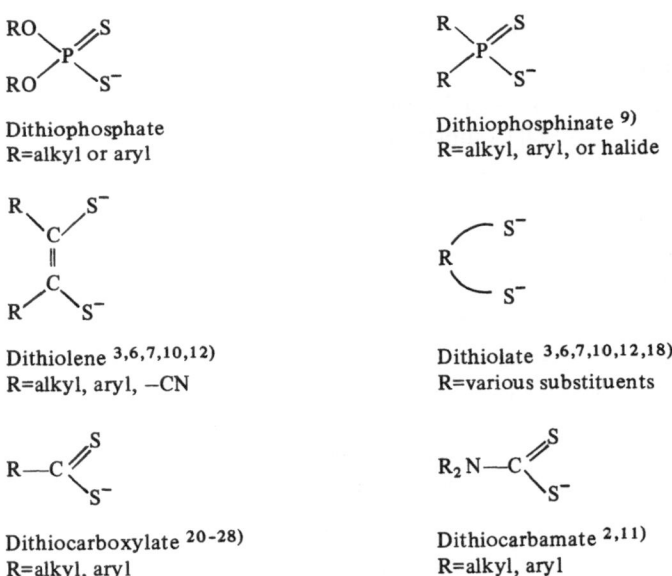

Dithiophosphate
R=alkyl or aryl

Dithiophosphinate [9]
R=alkyl, aryl, or halide

Dithiolene [3,6,7,10,12]
R=alkyl, aryl, −CN

Dithiolate [3,6,7,10,12,18]
R=various substituents

Dithiocarboxylate [20-28]
R=alkyl, aryl

Dithiocarbamate [2,11]
R=alkyl, aryl

Fig. 1. Examples of sulfur donor ligands. The selenium and mixed selenium-sulfur analogs of few of these ligands have been investigated

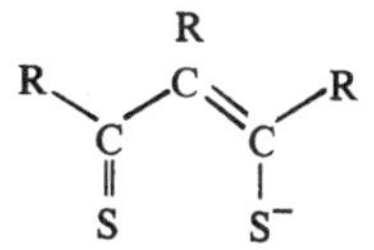

Dithio-β-diketonate [15,16]
R=alkyl, aryl, H

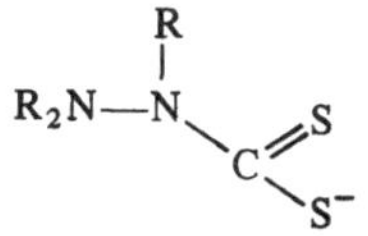

Dithiocarbazate [29-31]
R=alkyl, aryl

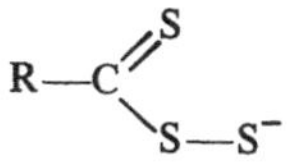

Perthiocarboxylate [32-35]
R=alkyl, aryl

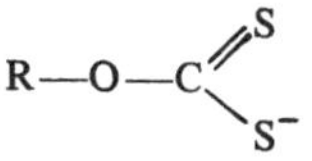

Xanthate [11]
R=alkyl

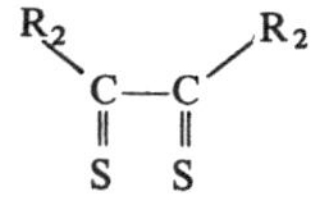

Dithiooxamide [1,37-40]
R=alkyl

R—S—CH$_2$—CH$_2$—S—R

Dithioethers [41-44]
R=alkyl, aryl and various
functional groups

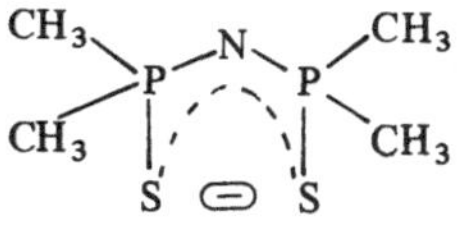

Imidotetramethyldithiodiphos-
phinate [46-49]

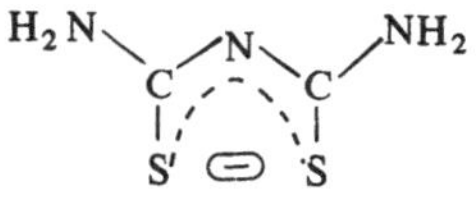

Dithiobiuretate [50-51]

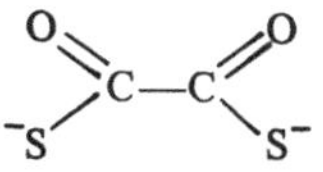

Dithiooxalate [1,52-55]

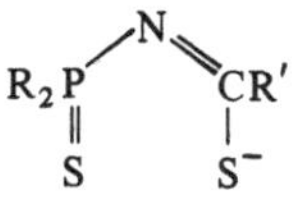

Phosphinothioylthioureas [56,57]
R=phenyl R'=alkyl

Metal complexes of dithiophosphoric acid, *1*, have found extensive application as oxidation inhibitors in various petroleum products. It is not surprising that many complexes are described in the patent literature [58-87] as well as

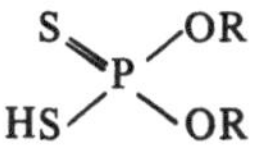

R = alkyl, aryl

1

literature concerning petroleum products [88-103].

Metal dithiophosphate complexes can be considered to be derivatives of the parent esters. Since World War II there has been wide-spread usage of dithiophosphate esters, *e.g.*, *2–4*, as pesticides and also related development of such compounds as war gases [104,105]. Dithiophosphate esters inhibit the action of several ester-splitting enzymes in living organisms.

$$C_2H_5O \diagdown P \diagup S \qquad S \diagdown P \diagup OC_2H_5$$
$$C_2H_5O \diagup \qquad {}^{\diagdown}S-CH_2-S \diagup \qquad {}^{\diagdown}OC_2H_5$$

2 (Trade name: ETHION)

$$CH_3O \diagdown P \diagup S$$
$$\qquad\qquad\qquad H$$
$$CH_3O \diagup \quad {}^{\diagdown}S-C-CO_2C_2H_5$$
$$\qquad\qquad\qquad |$$
$$\qquad\qquad H_2C-CO_2C_2H_5$$

3 (Trade name: MALATHION)

$$C_2H_5O \diagdown P \diagup S$$
$$C_2H_5O \diagup \quad {}^{\diagdown}S-CH_2-S-C_2H_5$$

4 (Trade name: THIMET)

They are singularly effective against cholinesterase which hydrolyzes the acetylcholine generated in myoneural junctions during the transmission of motor commands. In the absence of effective cholinesterase, acetylcholine accumulates and interferes with the coordination of muscle response. Such interference in the muscles of the vital organs produces serious symptoms and eventually death.

The widespread application and development of dithiophosphate esters have far outdistanced efforts to provide methods for the analysis of such materials. For ETHION *(2)* Graham [106] has shown that it is possible to analyze for the ester by hydrolyzing it in the presence of copper (II) ion and then determining the amount of the copper complex spectrophotometrically. A yellow copper complex similarly formed with MALATHION serves [107] as the basis for colorimetric determination of that pesticide. Further development of analytical methods for dithiophosphate esters will probably employ transition metal complexes to an appreciable extent. However, colorimetric methods will be of limited utility, since, as discussed later, the electronic spectra of metal dithiophosphates are not particularly sensitive to R-groups on the ligands. Development of dithiophosphate esters continues to be of interest [108-110] and additional commercial interest will develop for the metal complexes since they have been shown [111,112] to be biologically active, too.

Metal dithiophosphate complexes are involved in a wide variety of analytical methods for metals. Diethyldithiophosphoric acid reportedly [113-115] forms complexes with thirty-five elements, mainly metals, in various oxidation states which are useful for solvent extraction. No attempt is made here to detail all the applications to solvent extraction methods which have been described since these are summarized in the texts listed. Dithiophosphoric acids are also employed as flotation reagents for the recovery of metals from their solutions [116,117].

Although cognizant of the applications, realized and potential, of metal dithiophosphate complexes, we only describe the preparation, properties and reactions of reasonably well-characterized compounds here. This is so in view of the availability and reliability of the literature reviewed, not from a desire to neglect the contributions to dithiophosphate chemistry available in the rather voluminous patent literature.

Finally, a remark about nomenclature is in order. Throughout the literature the terms: O, O-dialkyldithiophosphate, O, O'-dialkyldithiophosphate, O, O-dialkylphosphorodithioate, O, O'-dialkyldithiophosphato and dialkyl-dithiophosphate, are used interchangeably for compounds containing the $(RO)_2PS_2$-group. Here, dithiophosphate alone refers to complexes containing the $(RO)_2PS_2$-group unless otherwise stated. Dithio- and diselenophosphate ligands are abbreviated. R-dtp and R-dsep, respectively, where R, the alkyl or aryl group, is specified.

II. Syntheses of Ligands and Complexes

1. Sulfur Derivatives

The reasonably well-established dithiophosphate complexes are listed in Table 1. Complexes alleged to be involved in colorimetric analytical methods, solvent extraction and flotation procedures are generally omitted unless they have also been characterized in the solid state.

Dithiophosphoric acids are readily prepared by the reaction [199-201]. When

$$8ROH + P_4S_{10} \longrightarrow 4(RO)_2P{\overset{\displaystyle S}{\underset{\displaystyle SH}{\diagup}}} + 2H_2S\uparrow$$

ROH is a low molecular weight alcohol, the alcohol serves as both reactant and solvent for the reaction. When ROH is a solid alcohol or phenol, the reaction can be conducted in a hydrocarbon solvent, *e.g.*, toluene, or by fusing [201] a mixture of ROH and P_4S_{10}. Only moderate heating of the reaction mixtures should be employed since temperatures much above 100 °C cause secondary reactions. Mass spectra of acids prepared by the above reaction usually indicate from about 1—5% impurities with molecular weights greater than the acids. The acids prepared by the alcoholysis of phosphorus(V) sulfide are frequently unstable to oxidative-hydrolytic action of the atmosphere; thus, immediate conversion of the acids to metal complexes is recommended. The acids can be stored under nitrogen for reasonable periods of time without extensive de-

Table 1. Metal dithiophosphate complexes

Complex		Ref.
V(ethyl-dtp)$_3$		118,119)
VO(ethyl-dtp)$_2$		120,121)
Cr(R-dtp)$_3$	R=methyl	122)
	ethyl	119,122,123,124,125,132)
	n-propyl	122)
	isopropyl	122)
	sec-butyl	122)
	isobutyl	122)
NbCl(OCH$_3$)$_2$(ethyl-dtp)$_2$		126)
NbBr(OCH$_3$)$_2$(ethyl-dtp)$_2$		126)
NbCl(OCH(CH$_3$)$_2$)$_2$(ethyl-dtp)$_2$		126)
NbCl(OC$_2$H$_5$)$_2$(ethyl-dtp)$_2$		126)
NbCl(OCH$_3$)$_2$(cyclohexyl-dtp)$_2$		126)
NbCl(OCH(CH$_3$)$_2$)$_2$(cyclohexyl-dtp)$_2$		126)
NbCl(OCH$_3$)$_2$(methyl-dtp)$_2$		126)
Nb(ethyl-dtp)$_4$		127)
Zr(ethyl-dtp)$_4$		127)
Mo$_2$O$_3$(ethyl-dtp)$_4$		128)
MoO(isopropyl-dtp)$_2$		129)
MoO(ethyl-dtp)$_2$		130)
Mo$_2$O$_3$(ethyl-dtp)$_4$		130,131)
Mo$_2$O$_3$(phenyl-dtp)$_4$		131)
Mo$_2$O$_3$(ethyl-dtp)$_4$(1,2-dichlorobenzene)$_2$		162)
Mn(CO)$_4$(methyl-dtp)		133)
Mn(CO)$_3$(pyridine)(methyl-dtp)		133)
Mn(CO)$_2$[P(OCH$_3$)$_3$]$_2$(methyl-dtp)		133)
Mn(CO)$_2$[P(CH$_3$)$_2$(C$_6$H$_5$)]$_2$(methyl-dtp)		133)
Mn(CO)$_3$(2,2'-bipyridine)(methyl-dtp)		133)
Mn(CO)$_3$(1,10-phenanthroline)(methyl-dtp)		133)
Mn(CO)$_4$(phenyl-dtp)		203)
Mn(CO)$_4$(ethyl-dtp)		203)
Mn(CO)$_4$(SOP(OC$_2$H$_5$)$_2$)		203)
Mn(CO)$_3$(phenyl-dtp)L	L=pyridine	203)
	L=2,2'-bipyridine	203)
	L=P(OC$_6$H$_5$)$_3$	203)
Fe(R-dtp)$_3$	R=methyl	134)
	ethyl	132,134,135,202)
	isopropyl	135,136)
	sec-butyl	135)
Ru(ethyl-dtp)$_3$		137)
Co(R-dtp)$_3$	R=methyl	134)
	ethyl	119,138)

Table 1 (continued)

Complex			Ref.
Rh(ethyl–dtp)$_3$			132,139)
Rh(ethyl–dtp)$_3$(triphenylphosphine)$_3$			140)
Ir(ethyl–dtp)$_3$			132,139,141)
H$_2$Ir(triphenylphosphine)$_2$(ethyl–dtp)			141)
H$_2$Ir(triphenylarsine)$_2$(ethyl–dtp)			141)
Ni(R–dtp)$_2$	R=methyl		112,142,259)
	ethyl		132,149,157,163-165,284)
	n-propyl		144)
	isopropyl		143)
	n-butyl		144)
	isobutyl		144)
	sec-butyl		144)
	cyclopentyl		145)
	cyclohexyl		112,146,147)
	2-phenylethyl		148)
	phenyl		145)
	2-chloroethyl		112)
	1,1-pentafluorophenylethyl		112)
Ni(R–dtp)$_2$L	R=methyl	L=2,2′-bipyridine	112,150,169)
	R=ethyl	L=2,2′-bipyridine	112,150,155)
	R=cyclohexyl	L=2,2′-bipyridine	112,150)
	R=isopropyl	L=2,2′-bipyridine	112,150)
	R=2-chloroethyl	L=2,2′-bipyridine	112)
	R=1,1-pentafluoro-		
	phenylethyl	L=2,2′-bipyridine	112)
	R=n-propyl	L=2,2′-bipyridine	150,156)
	R=n-butyl	L=2,2′-bipyridine	150)
	R=sec-butyl	L=2,2′-bipyridine	150)
	R=isobutyl	L=2,2′-bipyridine	150)
	R=methyl	L=1,10-phenanthroline	112)
	R=ethyl	L=1,10-phenanthroline	112,155,166)
	R=cyclohexyl	L=1,10-phenanthroline	112)
	R=isopropyl	L=1,10-phenanthroline	112)
	R=2-chloroethyl	L=1,10-phenanthroline	112)
	R=1,1-pentafluoro-		
	phenylethyl	L=1,10-phenanthroline	112)
	R=propyl	L=1,10-phenanthroline	156)
	R=ethyl	L=2,2′,2″-terpyridine	112)
	R=isopropyl	L=2,2′,2″-terpyridine	112)
	R=ethyl	L=di-n-butylamine	151)
	R=ethyl	L=diethylamine	151,154,155)
	R=ethyl	L=piperidine	152)

71

Table 1 (continued)

Complex			Ref.
	R=ethyl	L=2,9-dimethyl-1,10-phenanthroline	156)
	R=methyl	L=2,9-dimethyl-1,10-phenanthroline	156)
Ni(R–dtp)$_2$L$_2$	R=methyl	L=pyridine	112,144)
		L=γ-picoline	112)
	R=ethyl	L=pyridine	112,144,153,156,158, 167)
		L=α-picoline	112)
		L=n-butylamine	152,154)
	R=cyclohexyl	L=pyridine	112)
		L=γ-picoline	112)
	R=1,1-pentafluoro-phenylethyl	L=pyridine	112)
		L=γ-picoline	112)
	R=n-propyl	L=pyridine	144)
	R=isopropyl	L=pyridine	144)
		L=β-picoline	143)
		L=γ-picoline	143)
	R=n-butyl	L=pyridine	144)
	R=isobutyl	L=pyridine	144)
	R=sec-butyl	L=pyridine	144)

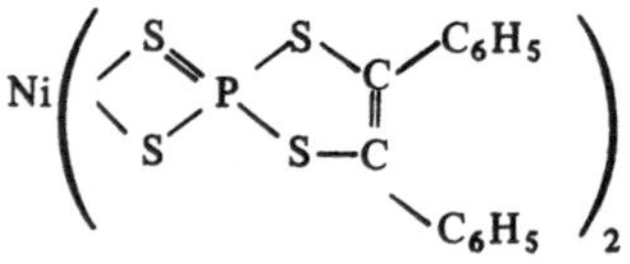

171)

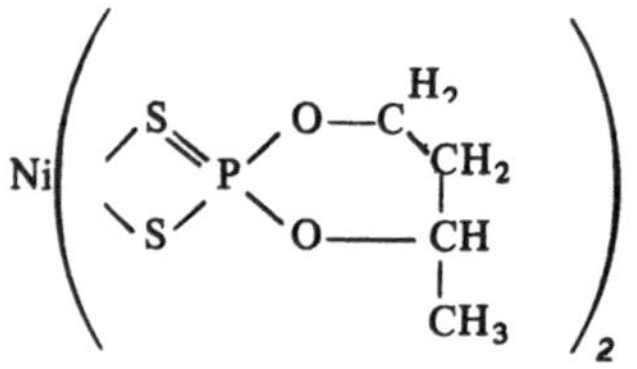

172)

Pd(methyl–dtp)$_2$		142)
Pd(ethyl–dtp)$_2$		132,284)
Pt(ethyl–dtp)$_2$		132,284)
Cu(k–dtp)$_2$	R=ethyl	121,159-161)

Table 1 (continued)

Complex		Ref.
[Cu(isopropyl-dtp)]$_4$		173,174)
Cu(methyl-dtp)		107)
{[Au(isopropyl-dtp)]$_2$}$_n$		175,176)
Zn(4-methylpentyl-2-dtp)$_2$		190)
Zn(ethyl-dtp)$_2$		177,185)
Zn(R-dtp)$_2$	R=butyl, isobutyl, sec-butyl, cyclohexyl	185,189)
Zn(isopropyl-dtp)$_2$		178,184,185,188)
Zn(methyl-dtp)$_2$		142,193)
Zn$_4$O(R-dtp)$_6$	R=ispropyl, n-butyl	182)
Zn$_2$(OH)(R-dtp)$_3$	R=n-butyl, isopropyl, isobutyl, propyl,ethyl	173,182,183,185, 186,191)
Zn(R-dtp)$_2$	R=n-amyl, 2-pentyl, 3-pentyl	187)
Cd(isopropyl-dtp)$_2$		178)
Cd(methyl-dtp)$_2$		142)
Hg(isopropyl-dtp)$_2$		173,179)
Hg(ethyl-dtp)$_2$		220,221)
Hg(r-dtp)$_2$	R=alkyl	200,213,220,221)
C$_2$H$_5$Hg(ethyl-dtp)		221)
Pb(isopropyl-dtp)$_2$		180,181)
Pb(methyl-dtp)$_2$		142)
Pb(ethyl-dtp)$_2$		200)
In(methyl-dtp)$_3$		142)
In(ethyl-dtp)$_3$		132,192)
Tl(methyl-dtp)		142,193)
Tl(ethyl-dtp)		193)
Tl(p-chlorophenyl-dtp)		193)
(CH$_3$)$_2$Tl(methyl-dtp)		193)
(CH$_3$)$_2$Tl(ehtyl-dtp)		193)
(CH$_3$)$_2$Tl(p-chlorophenyl-dtp)		193)
(C$_2$H$_5$)$_2$Tl(methyl-dtp)		193)
(C$_2$H$_5$)$_2$Tl(ethyl-dtp)		193)
As(ethyl-dtp)$_3$		202)
Sb(ethyl-dtp)$_3$		202)
Sb(R-dtp)$_3$	R=alkyl	100)
Bi(ethyl-dtp)$_3$		142,202)
Se(methyl-dtp)$_2$		197)
Se(methyl-dtp)$_2$		194,195,196,197)
Te(methyl-dtp)$_2$		196,197,198)
Te(ethyl-dtp)$_2$		197)
(methyl-dtp)Sn(C$_6$H$_5$)$_3$		111)

composition. It is noted that the acids and their complexes should be handled with caution since they are all potentially biologically active.

Details of the preparations of numerous dithiophosphoric acids have been reported [64,66,68,70,79,84,85,199-201,204-211]. Ammonium, alkali and alkaline metal salts of the acids can be obtained by neutralization of the acids with hydroxides or carbonates [200,204,212-215]. The sodium salts are probably best prepared by the procedure described by Makens, Vaughan and Chelberg [213].

An alternate synthesis of dithiophosphoric acids, which is also applicable to the preparation of mixed acids *(2, 3)* [200,214,216,217], described by Malatesta [218], involves the reaction:

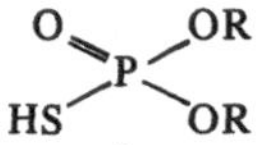

$$
\begin{array}{ccc}
O{\diagdown}\!\!\!\!\!\!\underset{HS}{\overset{}{}}\!\!>\!\!P\!\!<\!\!\overset{OR}{\underset{OR}{}} & \quad & S{\diagdown}\!\!\!\!\!\!\underset{HSe}{\overset{}{}}\!\!>\!\!P\!\!<\!\!\overset{OR}{\underset{OR}{}} \\
\mathbf{2} & & \mathbf{3}
\end{array}
$$

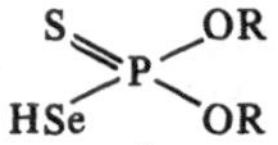

$$S = PCl_3 + 2NaOR \longrightarrow \overset{\overset{S}{\|}}{Cl-P(OR)_2} + 2NaCl$$

$$\mathbf{4}$$

The product, *(4)*, can subsequently be converted to the potassium salt of the respective acid by reaction with KOH, KSH, KSeH [200]. The same material *(4)*, can also be obtained from the following reaction [201]:

$$2(RO)_2\overset{\overset{S}{\|}}{P}-SH + 3Cl_2 \longrightarrow 2(RO)_2\overset{\overset{S}{\|}}{P}-Cl + S_2Cl_2 + 2HCl$$

$$\mathbf{4}$$

The preparations of mixed phosphinic acids, *e.g.*, $(C_6H_5)_2PSSeH$ [219], involve similar types of reactions.

Oxidation of dithiophosphoric acids can give rise to a variety of products. The chlorination of dithiophosphoric acids gives rise to products such as *(4)*, but other products [218] can also be obtained. Oxidation of the acids or their salts with iodine yields *bis*-(O, O'-dialkyldithiophosphoryl)disulfides,

$(RO)_2\overset{\overset{S}{\|}}{P}-S-S-\overset{\overset{S}{\|}}{P}(OR)_2$. A number of these products and others from reactions of dithiophosphoric acids with oxidants are listed in Table 2 since they are some of the impurities to be anticipated. Thiophosphoryl (P=S) compounds are rapidly, quantitatively, and stereospecifically converted to phosphoryl (P=O) compounds by organic peroxy acids under mild conditions [222]. The reactions of peroxy acids and dithiophosphoric acids and salts have apparently not been characterized.

Table 2. Derivatives of dithiophosphoric acids

Compound		Ref.
$S[SP(OC_2H_5)_2]_2$		218)
$S_2[SP(OC_2H_5)_2]_2$		213,218)
$S_3[SP(OC_2H_5)_2]_2$		218)
$S_4[SP(OC_2H_5)_2]_2$		218)
$S_2[SP(OCH_3)_2]_2$		107)
$S_2[SP(OR)_2]_2$	R=alkyl	220)
$S_2[SP(O-iso\ C_3H_7)_2]_2$		213,226,227)

Schrauzer, Mayweg and Heinrich [171] have reported the preparation of the complex *5* from the reaction of aqueous nickel(II) chloride with a reaction mixture of benzoin in dioxane and phosphorus(V) sulfide. The parent acid and its salts have not been characterized. The only other closed ring complex, *6*, was isolated in low yield (3%) subsequent to the reaction of nickel (II) acetate with the mixture resulting from the reaction of 1,3-propanediol with phosphorus(V) sulfide. As for *5*, the parent acid and its salts were not

5   6

characterized [172]. Presumably, concomitant polymerization reactions of some diols with phosphorus(V) sulfide account for the low yields of complexes with cyclic P structures. The potassium salt of *cis*-cyclohexane-1,2-diol-dithiophosphoric acid can be obtained in 62% yield [172a].

The preparation of metal dithiophosphate complexes usually involves the reaction of metal halides or acetates with dithiophosphoric acids or their salts. The metal complexes are generally purified by repeated fractional crystallization from halocarbon solvents such as chloroform. The reactions of mixtures of alcohols and alcohols and phenols with phosphorus(V) sulfide allegedly,

e.g., Ref. [212], give rise to acids of the type $HSP(OR)(OR')$. Few, if any, complexes with R≠R' have been definitely established. Since the properties of dithiophosphate complexes with a given metal ion are so similar, separation of complexes with R≠R' is expected to be very difficult. Thin-layer and paper chromatographic separations [101-103] offer some hope for the isolation of such complexes.

Preparations of complexes where water is present should be considered with caution since several reactions are known to exhibit a pH dependence. In slightly basic solutions zinc(II) complexes of the type

$$OZn_4[S_2P(OR)_2]_6$$

are obtained whereas acidic solutions yield the normal *bis*-complex [182]. Jφr-gensen [132] noted that the pure indium(III) and bismuth(III) complexes are precipitated from acidic solutions.

Reactions of dithiophosphoric acids and their salts with complexes are also employed to prepare dithiophosphate complexes, *e.g.*,

Ref. [111]: $\{(C_6H_5)_3Sn\}_2O + 2HSP(OCH_3)_2 \rightarrow 2(C_6H_5)_3Sn-S-\overset{\overset{\displaystyle S}{\|}}{P}(OCH_3)_2 + H_2O$

Ref. [113,203]: $Mn(CO)_5Br + HSP(OR)_2 \rightarrow HBr + CO + Mn(CO)_5(R-dtp)$

Table 3. Mixed dichalcogenophosphate complexes

Compound		Ref.
Ni[SSeP(OC$_6$H$_5$)$_2$]$_2$		216)
Pb[SSeP(OC$_6$H$_5$)$_2$]$_2$		216)
Metal[OSP(O-nC$_4$H$_9$)$_2$]$_n$	n=oxidation state of metal ion	224)
Ag[OSP(OR)$_2$]	R=propyl, isopropyl, n-butyl	217)
Co[OSP(OC$_2$H$_5$)$_2$]$_2$		223)
Co[OSeP(OC$_2$H$_5$)$_2$]$_2$		223,356)
SnCl$_4$ · 2L	L= HS$-\overset{\overset{\displaystyle O}{\|}}{P}$(O$n-C_4H_9$)$_2$	225)
	L= O=P(OC$_2$H$_5$)$_2$(SC$_2$H$_5$)	225)
	L= HS$-\overset{\overset{\displaystyle O}{\|}}{P}$(CH$_3$)(OCH$_3$)	225)
SnBr$_4$ · 2L	L= HS$-\overset{\overset{\displaystyle O}{\|}}{P}$(CH$_3$)(OCH$_3$)	225)
Metal [SSeP(OC$_6$H$_5$)$_2$]$_n$	n = oxidation state of metal	355)
M[OSP(OC$_2$H$_5$)$_2$]$_3$	M= V, Cr, Fe	372)

Ref. [132]: $Na_3RhCl_6 + 3NH_4(ethyl\text{-}dtp) \rightarrow Rh(ethyl\text{-}dtp)_3 + 3NaCl + 3NH_4Cl$

$$\overset{\overset{\displaystyle S}{\|}}{}$$

Ref. [228]: $K_3Co(CO_3)_3 + 3HS\text{-}P(OR)_2 \rightarrow Co[S_2P(OR)_2]_3 + 3CO_2 + 3KOH$

A variety of mixed dichalcogenophosphate complexes have been reported and these are listed in Table 3. Mixed dichalcogenophosphoric acids exhibit tautomeric equilibria of the type [214,224];

$$(RO)_2\overset{\overset{\displaystyle X}{\|}}{P}\text{-}SH \rightleftharpoons (RO)_2\overset{\overset{\displaystyle S}{\|}}{P}\text{-}XH \qquad X=O, Se$$

Thiol   Thiono

When the ligand anions are monodentate, the donor coordination site employed must be established.

2. Selenium Derivatives

Diselenophosphate complexes are prepared from the interaction of metal salts and complexes with appropriate diselenophosphoric acid or its salt. The acids are obtained from the reaction of phosphorus(V) selenide with alcohols [229]. The preparation of phosphorus(V) selenide and its reactions with alcohols [229] and amines [230] have been described and a variety of complexes reported (Table 4). The biological activity of these compounds does not seem to have described but the exercise of extreme caution when handling these materials is recommended. Zingaro and his coworkers [229-232] have thoroughly characterized the thermal and spectroscopic properties of a number of compounds.

Table 4. Metal diselenophosphate complexes

Complex	Ref.
$Cr(ethyl\text{-}dsep)_3$	229,233,234)
$Cr(n\text{-}propyl\text{-}dsep)_3$	229)
$Rh(ethyl\text{-}dsep)_3$	233)
$Ir(ethyl\text{-}dsep)_3$	232,233)
$Ni(ethyl\text{-}dsep)_2$	232,233)
$Co(ethyl\text{-}dsep)_3$	232,234)
$Ni(ethyl\text{-}dsep)_2(pyridine)_2$	232)
$K(ethyl\text{-}dsep)$	229,232)
$Ag(ethyl\text{-}dsep)$	232)
$Cu(ethyl\text{-}dsep)$	232)

77

Table 4 (continued)

Complex		Ref.
Zn(ethyl-dsep)$_2$		232)
Cd(ethyl-dsep)$_2$		232)
Tl(ethyl-dsep)$_2$		231)
Pb(ethyl-dsep)$_2$		231)
Sn(ethyl-dsep)$_2$		231)
As(ethyl-dsep)$_3$		231)
Sb(ethyl-dsep)$_3$		231)
Bi(ethyl-dsep)$_3$		231)
In(ethyl-dsep)$_3$		231,233)
RNH$_3^{+-}$Se$_2$P(NHR')$_2$	R=R'=hexyl; R=R'=butyl and isobutyl	230)
KSe$_2$P(NHR)$_2$	R=butyl	230)
(RO)$_4$P$_2$Se$_4$	R=cyclohexyl or isopropyl	229)
(RO)$_4$P$_2$Se$_5$	R=ethyl	229)
Metal[S$_2$P(OR)$_2$]$_n$	n = oxidation state of metal R=C$_6$H$_5$, CH$_3$, n-C$_3$H$_1$, iso-C$_3$H$_1$ n-C$_4$H$_9$, C$_2$H$_5$	351)

III. Structural Data

1. Non-Transition Metal Compounds

In this and the following section the crystal structures of dithiophosphate compounds are summarized and compared to those of related compounds. As is obvious from Tables 5 and 6, dithiophosphate compounds comprise a fruitful area for structural research. Additional data for comparative purposes may be found in Corbridge's [235] review of the structural chemistry of phosphorus compounds.

The P=S bond distance is frequently found to lie in the range 1.85—1.96 Å and the P—S bond distance within the range 2.08—2.19Å. The P=O bond distance generally is found in the range 1.39—1.54Å and the P—O bond distance within the range 1.56—1.64Å [235]. Bond distances calculated from diffraction data are frequently assigned and charge localization discussed in terms of the preceding criteria, e.g., for K[S$_2$P(OCH$_3$)$_2$] [236] it has been inferred that most of the negative charge on the ligand anion resides on the sulfur atoms.

Two significant general results of the structural data thus far reported are: (1) dithiophosphate, S$_2$P(OR)$_2$, groups can behave as mono- and bidentate ligands and (2) substitution of one R group for another usually leads to changes

Table 5. Structures of non-transition metal dithiophosphates and related compounds

Compound	Comments	Ref.
$K[S_2P(OCH_3)_2]$	P–S 1.96 Å P–O 1.64 Å O–C 1.58 Å PS_2O_2 tetrahedra K^+ ions in irregular 8-coordination	236)
$[S_2P(O-iC_3H_7)_2]_2$		227)

P–O 1.56 Å
P–O–C 121°

$H_3PO_4 \cdot 0.5H_2O$	a) P–O 1.503,1.554,1.545 and 1.561 Å b) P–O 1.477,1.542,1.557 and 1.554 Å approximate PO_4 tetrahedra	237)

	P–S (Å)	P=S (Å)	
P_4S_3	2.090	–	238)
P_4S_5	2.08–2.19	1.94	
P_4S_7	2.08	1.95	
P_4S_{10}	2.08	1.95	
$\beta\text{-}P_4S_3I_2$	2.221,2.091,2.2097	–	239)

Compound	Comments	Ref.
	P–Cl 2.059 Å P–S 2.051 Å P=S 1.910 Å	240)
Triethylammonium Uridine 2′,3′-O–O-cyclo- phosphorothioate	P=S 1.946 Å P–O 1.48 Å	241)
$(NH_4)_2[PO_2S(NH_2)]$	PO_2SN tetrahedra P–O 1.522 and 1.544 Å P–S 2.051 Å P–N 1.697 Å	242)
$OZn_4[S_2P(OR)_2]_6$	OZn_4 tetrahedra Zn–Zn bridging R-dtp Structure analogous to that of basic beryllium acetate	182)

Table 5 (continued)

Compound	Comments	Ref.
$(Zn[S_2P(OC_2H_5)_2]_2)_n$	Distorted ZnS_4 tetrahedra Zn–S 2.345,2.337,2.351 and 2.401 Å One dithiophosphate bidentate One dithiophosphate bridges Zn atoms Infinite polymers P–S 1.973–2.001 Å; P–O 1.56–1.62 Å	177)
$\{Zn[S_2P(O\text{-iso }C_3H_7)_2]_2\}_2$	Distorted ZnS_4 tetrahedra, binuclear Zn–S distances range from 2.302 to 2.409 Å Zn–Zn distance 4.108 Å One dithiophosphate bidentate One dithiophosphate bridging P–S average 1.970 Å P–O average 1.58 Å	178)
$\{Cd[S_2P(O\text{-iso }C_3H_7)_2]_2\}_2$	Structure similar to Zn compound above Cd–Cd distance 4.059 Å P–S average 1.965 Å P–O average 1.58 Å Cd–S distances range from 2.486–2.590 Å	178)
$Hg[S_2P(O\text{-iso }C_3H_7)_2]_2$	One dithiophosphate bidentate One dithiophosphate bridging Helical chain polymer Mercury five-coordinate Hg–S 2.388,2.391,2.748,2.888 and 3.408 Å P–S average 2.01 Å (associated with shorter Hg–S bonds) P–S average 1.94 Å (associated with longer Hg–S interactions) P–O average 1.62 Å for nonbridging ligands P–O average 1.54 Å for bridging ligands	179)
$Hg[S_2CN(CH_3)_2]_2I_2$	Hg–S 2.651 and 2.882 Å	256)

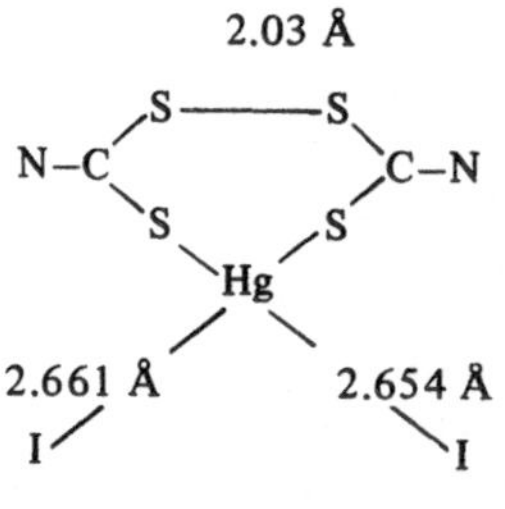

2.03 Å

2.661 Å   2.654 Å

Table 5 (continued)

Compound	Comments	Ref.
$[Zn(S_2CN(C_2H_5)_2)_2]_2$	ZnS_5 distorted tetragonal pyramid, binuclear Zn–Zn 3.546 Å Zn–S 2.443, 2.355, 2.331, 2.815 and 2.383 Å	[243]
$Zn[S_2CN(CH_3)_2]_2 \cdot C_5H_5N$	$ZnNS_4$, distorted trigonal bipyramid Zn–S two at 2.60 Å Zn–S two at 2.33 Å Zn–N 2.08 Å	[244]
$\{Cd[S_2CN(C_2H_5)_2]_2\}_2$	CdS_5 distorted tetragonal pyramid, binuclear Cd–S distances range from 2.536 to 2.800 Å	[245]
$\{Zn[SOP(n\text{-}C_4H_9)_2]_2\}_n$	Polymeric chains Isostructural to Co(II) complex Distorted ZnO_2S_2 tetrahedra	[246]
$\{Zn[S_2P(C_2H_5)_2]_2\}_2$	Distorted ZnS_4 tetrahedra One dithiophosphinate bridging One dithiophosphinate bidentate Zn–S 2.302–2.382 Å	[247]
$Zn[Se_2CN(C_2H_5)_2]_2$	Isostructural with sulfur analogue Zn–Se 2.435, 2.568, 3.003, 2.446 and 2.492 Å	[248]
$Pb[S_2P(O\text{-iso } C_3H_7)_2]_2$	Polymeric chains Distorted PbS_6 octahedron Pb–S two at 2.766, two at 3.01 and two at 3.20 Å Nearly planar $Pb(S_2P)_2$ grouping P–S 1.944, 1.958, 1.982 and 2.00 Å P–O 1.556, 1.582, 1.584 and 1.584 Å	[180, 181]
$In[S_2P(OC_2H_5)_2]_3$	Distorted InS_6 octahedron P–O average 1.60 Å Two P–S for each ligand In–S average 2.608 Å P–S average 1.995 Å	[192]
$(C_6H_5)_2P\overset{\diagup S}{\underset{}{\text{—OCH}_3}}$	P=S 1.94 Å	[249]
$(C_6H_5)_2P\overset{\diagup Se}{\underset{}{\text{—OCH}_3}}$	P=Se 2.09 Å	
$Te[S_2P(OCH_3)_2]_2$	Approximate TeS_4 square planar coordination P–S 1.92 and 2.09 Å P–O 1.57 and 1.59 Å Te–S 2.44 Å	[198]

Table 5 (continued)

Compound	Comments	Ref.
$(C_2H_5)_2P\overset{S}{\underset{}{\diagup}}Se-Se\overset{S}{\underset{}{\diagdown}}P(C_2H_5)_2$	P=S 1.93 Å P–Se 2.28 Å Se–Se 2.33 Å	250)

Table 6. Structures of transition metal dithiophosphates and related compounds

Compound	Structure	Ref.
$VO[S_2P(CH_3)_2]_2$	C_{2v} moleculare symmetry Angle between OV and SVS bisector of $105°$	251)
$V[S_2P(OC_2H_5)_2]_3$	V–S average 2.45 ± 0.02 Å VS_6 trigonally distorted octahedron P–S average 1.98 ± 0.01 Å VS_2P unit planar	118)
$V\left(\begin{array}{c} S-C-C_6H_5 \\ S-C-C_6H_5 \end{array}\right)_3$	VS_6 trigonal prismatic V–S 2.338 Å C=C 1.41 Å C–S average 1.69 Å	253)
$V\left(\begin{array}{c} S \\ S \end{array}C-CH_2-C_6H_5\right)_4$	VS_8 dodecahedron V–S 2.524–2.470 Å	23,254)
$BaVS_3$	Polymeric Distorted VS_6 octahedra V–S 2.385 Å	255)
$Mo_2O_3[S_2P(OC_2H_5)_2]_4 \cdot 2$ (dichlorobenzene structure)	Mo–S 2.496 Å mean Mo=O 1.647 Å Mo–O 1.863 Å P–S 1.95–2.02Å P–O 1.55 Å MoO_2S_4 octahedra Mo–O–Mo linear	162)
$Fe[SP(CH_3)_2N(CH_3)_2PS]_2$	FeS_4 tetrahedra Not isomorphous corresponding Co(II) and Ni(II) species Fe–S 2.339–2.380 Å P–S 2.020 Å P–C 1.806 Å P–N 1.591 Å	48)

Table 6 (continued)

Compound	Structure	Ref.
Ni[SP(CH$_3$)$_2$N(CH$_3$)$_2$PS]$_2$	NiS$_4$ tetrahedra Ni−S 2.282 Å P−C 1.825 Å P−S 2.023 Å P−N 1.580 Å	47)
Ni[S$_2$P(OCH$_3$)$_2$]$_2$ · 1,10-phenanthroline	NiS$_4$N$_2$ octahedra Ni−S 2.47−2.52 Å Ni−N 2.08, 2.09 Å P−S 1.97 Å P−O 1.56−1.59 Å	169,170)
Ni[S$_2$P(OCH$_3$)$_2$]$_2$ · 2,9-dimethyl-1,10-phenthroline	NiS$_3$N$_2$ Ni−N 1.97, 2.03 Å Ni−S 2.42, 2.30, 2.58 Å P−S 1.94, 1.96, 1.91, 1.97 Å P−O 1.56−1.60 Å	168,170)
Ni[S$_2$P(C$_2$H$_5$)$_2$] · quinoline	NiS$_4$N Ni−N 2.06 Å Ni−S 2.39−2.43 Å P−S 1.99−2.01 Å P−O 1.77−1.82 Å	170)
Ni[S$_2$P(OC$_2$H$_5$)$_2$]$_2$ · 1,10-phenanthroline	NiS$_4$N$_2$ octahedra Ni−S 2.49−2.509 Å Ni−N 2.104 Å P−S 1.959−1.972 Å P−O 1.595−1.621 Å	166)
Ni[S$_2$P(OC$_2$H$_5$)$_2$]$_2$	NiS$_4$ planar Ni−S 2.21 Å P−S 1.95−1.98 Å P−O 1.63 Å	163-165)
Ni[S$_2$P(OC$_2$H$_5$)$_2$]$_2$ · 2 pyridine	trans-NiS$_4$N$_2$ octahedra Ni−S 2.49 Å P−S 1.98 Å P−O 1.58 Å Ni−N 2.11 Å	167)
Ni[S$_2$P(C$_6$H$_5$)$_2$]$_2$ · 2 pyridine	trans-NiS$_4$N$_2$ octahedra Ni−S 2.50 Å P−S 2.00 Å P−C 1.80 Å Ni−N 2.08 Å	257)

Table 6 (continued)

Compound	Structure	Ref.
$Ni[S_2P(C_6H_5)_2]_2$	NiS$_4$ planar Ni–S 2.24Å P–S 2.01 Å P–C 1.78 Å	[258]
$Ni[S_2P(OCH_3)_2]_2$	NiS$_4$ planar Ni–S 2.22 Å P–S 1.98 Å P–O 1.56 Å	[259]
$Ni[S_2P(OCH_3)_2]_2 \cdot 2,2'$-dipyridyl	NiS$_4$N$_2$ octahedral Ni–S 2.48–2.52 Å Ni–N 2.07 Å	[169,260]
$Ni[S_2P(CH_3)_2]_2$	NiS$_4$ planar Ni–S 2.229–2.242 Å P–S 1.991–2.018 Å P–C 1.80–1.83 Å	[261]
$Ni[S_2P(C_2H_5)_2]_2$	NiS$_4$ planar Ni–S 2.22 Å P–S 2.00 Å P–C 1.84 Å	[262]
Ni(S–S–benzo)$_2$	NiS$_4$ planar Ni–S 2.147 Å	[263]
Ni(S–C(–NH$_2$)=N–NH$_2$)$_2$	red *trans*-NiS$_2$N$_2$ planar Ni–S 2.155 Å Ni–N 1.911 Å C–S 1.746 Å	[264]
Ni(S–S–C–NH$_2$)$_2$	NiS$_4$ planar Ni–S 2.21 Å C–S 1.69 Å C–N 1.37 Å	[265]
Ni(S–S–C–N(C$_2$H$_5$)$_2$)$_2$	NiS$_4$ planar Ni–S 2.20 Å C–S 1.71 Å C–N 1.33 Å	[266]
Ni(S–S–C–N(nC$_3$H$_7$)$_2$)$_2$	NiS$_4$ planar Ni–S 2.197–2.209 Å C–S 1.722 Å C–N 1.33 Å	[267]

Table 6 (continued)

Compound	Structure	Ref.
$Cu_4[S_2P(O\text{-iso }C_3H_7)_2]_4$	Cu_4 tetrahedron P–O 1.56 Å Bridging $S_2P(OC_2H_5)_2$	174)
$[\{Au[S_2P(O\text{-iso }C_3H_7)_2]\}_2]_n$	Linear chains of Au–Au bond between [Au dtp]$_2$ dimers Au–S 2.28 Å	175)

in structure, *e.g.*, compare Zn(ethyl-dtp)$_2$ and Zn(isopropyl-dtp)$_2$. Additional structural data are needed in order to thoroughly document the changes brought about by R group variation.

The P–S bond distances found for dithiophosphate complexes are fairly sensitive to the environment the sulfur atoms are located in — the more tightly bound to a given metal ion the sulfur is, the longer the P–S bond distance. The P–O bond distances appear to reflect stereochemistry rather than the strength of metal-sulfur interactions. The OPO and particularly the SPS bond angles vary appreciably indicating variable hybridization of the phosphorus. Huheey and his co-workers [268] have developed expressions for determining the s-character of each of the four hybrid orbitals of the central phosphorus atom and indicated that the s-character of the bond is not necessarily the dominant factor in determining the magnitude of phosphorus-heavy metal spin-spin coupling constants. The participation of phosphorus d-orbitals in bonding in dithiophosphate complexes, although expected [269-280], is difficult to assess from the structural and spectroscopic information now available.

For Te[S$_2$P(OCH$_3$)$_2$]$_2$ the ligand has been found to be monodentate [198]. With zinc(II) and cadmium(II) complexes the ligand, O, O'-di-isopropyl-dithiophosphate, has been shown to be both bidentate and bridging [178]. The irregular pentagonal bipyramidal structure [180,181] of Pb[S$_2$P(O-iso C$_3$H$_7$)$_2$]$_2$ has been accounted for in terms of the Gillespie-Nyholm valence-shell electron-pair repulsion (VSEPR) model.

2. Transition Metal Complexes

Table 6 collects the structures of transition metal dithiophosphate complexes along with comparative data for related compounds. As is readily noted, by far the most structural work is for nickel(II) complexes and their adducts.

Generally, complexes containing the NiS_4 chromophore are diamagnetic and square planar. *Bis*(imidotetramethyldithiodiphosphino-S,S)nickel(II), $Ni[SP(CH_3)_2 N(CH_3)_2 PS]_2$, however, is tetrahedral [47]. This is an anomaly which cannot be explained by steric crowding or an abnormally weak ligand field. Nickel(II) dithiophosphates, dithiophosphinates and other sulfur donor ligands adopt the square planar geometry. The addition of pyridine to nickel(II) dithiophosphates results in the formation of green, paramagnetic *trans*-octahedral complexes. Marked structural changes accompany adduct formation. The S–Ni–S angle in $Ni(ethyl-dtp)_2$ is decreased from 88 to 81.7° in $Ni(ethyl-dtp) . 2$ pyridine and the S-P-S angle is increased from 103° in $Ni(ethyl-dtp)_2$ to 110.4° in the pyridine adduct [167]. The nickel-nitrogen bond distance, 2.11Å, in $Ni(ethyl-dtp)_2 \cdot 2$ pyridine is normal for a neutral nitrogen ligand. However, adduct formation results in the increase of the Ni–S distance. The Ni–S bond distance is 2.21Å in $Ni(ethyl-dtp)_2$ while it is 2.49Å in the *bis*-pyridine adduct.

With sterically crowded adduct ligands distorted square pyramidal structures result. The 2,9-dimenthyl-1,10-phenanthroline adduct of $Ni[S_2P(OCH_3)_2]_2$ exhibits such a geometry [108], the Ni–S bond lengths being greater ($\sim 0.1-0.3$Å) than the parent molecule [259]. Generally, the five-coordinate adducts of $Ni(R-dtp)_2$ complexes are unstable but, in favorable instances, their existence can be detected [146-148,151-155]. With bidentate ligands, *e.g.*, 1,10-phenanthroline, green *cis*-octahedral NiS_4N_2 complexes are obtained. Again, adduct formation results in an increase of Ni–S bond distances on the order of 0.1–0.3Å. It is worthwhile pointing out that the species isolated in the solid state are not necessarily those present in solution. In solution a 1 : 1 adduct of a $Ni(R-dtp)_2$ complex with a bidentate ligand can assume *cis*-octahedral as well as square pyramidal five-co-ordinate geometries. Consideration of the equilibria possible in solutions of donor molecules and $Ni(R-dtp)_2$ complexes necessitates the use of caution when discussing solution species.

IV. Properties and Reactions

1. General Properties

Many dithiophosphate complexes have been found to be associated in solution. $Zn(ethyl-dtp)_2$ is monomeric in chloroform solution but polymerized in the solid state [177]). Association studies [178,281]) have shown that zinc, mercury and lead di-isopropyldithiophosphate complexes undergo monomer ⇌ dimer equilibrium in benzene whereas the corresponding cadmium complex is dimeric in concentrations above 0.005 gm/ml. In the solid state the zinc and cadmium complexes are dimeric [178]) whereas the mercury [179]) and lead [180,181]) complexes are polymeric. Dakternicks and Graddon [185]) have characterized several $Zn(R-dtp)_2$ complexes and their pyridine adducts in benzene. Pyridine was shown to break down the molecular association and the thermodynamics of depolymerization and adduct formation were investigated [282]). Association data for dialkylthallium and thallium(I) as well as $Zn[S_2P(OCH_3)_2]_2$ have also been reported [193]). Crystalline $\{[AuS_2P(O-isoC_3H_7)_2]\}_n$ is comprised of an indefinite one dimensional array of [$AuS_2P(O-isoC_3H_7)]_2$ units; however, vapor pressure osmometry molecular weight data in benzene at 37 °C indicate that at weight-per-cent concentrations of 4% or less these chains dissociate into dimeric units [175]). A number of barium R-dtp salts form aggregates of up to ten monomers, the association steps proceeding with equal ease [281]). Dimeric and polymeric structures are found for many dithiophosphinates [247]), thiophosphinates [246]) and phosphinates [283]) of metals; in the latter, polymerization is probably a consequence of the smaller chelate "bite" which is inadequate for the formation of four-membered chelate rings.

In the preceding section (III. 2) it was noted that charge localization in dithiophosphate complexes is frequently inferred from bond distance data. Unlike related dithiolene ligands, extensive conjugation of dithiophosphate sulfur donor sites with other parts of the molecules does not occur. This is evident from the electronic spectra of chromium(III) complexes discussed in Section V. 1. However, replacement of –OR groups on phosphorus with other groups does give rise to variations of electronic structural properties [120,123,252,284]). In order to provide a qualitative assessment of the charge distributions, and, hence, ligand field strengths of chalcogenophosphate ligands, charge distributions were calculated using Sanderson's approach [285-288]). The basis of the calculations is the electronegativity equalization principle which states: when two atoms of differing electronegativity unite to form a molecule, the electronegativity of each atom adjusts to an equal intermediate value. Sanderson has shown [285-287]) that electronegativity and partial charge are linearly related. The partial charge is calculated using the expression:

$$PC_i = \frac{S_m - S_{E_i}}{S_{E_i} \rightarrow E_{\pm i}}$$

where PC_i is the partial charge of element i, S_m is the geometric mean of the electronegativities of all the elements in the molecule, S_{E_i} is the Sanderson electronegativity of element i, and $S_{E_i \rightarrow E_{\pm i}}$ is the Sanderson electronegativity of the element if it has gained or lost an electron. The results of the calculations

Table 7. Sanderson charge distributions for dichalcogenophosphates partial charge

Compound	Metal ion	P	S	Se	O	F	C	H
Ag_3PO_4	+0.364	+0.116	—	—	-0.301	—	—	—
Ag_3PS_4	+0.237	+0.005	-0.178	—	—	—	—	—
Ag_3PSe_4	+0.254	+0.020	—	-0.195	—	—	—	—
$AgS_2(OCH_3)_2$	+0.358	+0.111	-0.083	—	-0.305	—	-0.007	+0.054
$Ni[S_2P(OCH_3)_2]_2$	+0.583	+0.136	-0.060	—	-0.285	—	+0.016	+0.078
$Cu[S_2P(OCH_3)_2]_2$	+0.385	+0.146	-0.051	—	-0.277	—	+0.026	+0.088
$AgSe_2P(OCH_3)_2$	+0.363	+0.115	—	-0.110	-0.301	—	-0.003	+0.058
$Ni[Se_2P(OCH_3)_2]_2$	+0.556	+0.122	—	-0.104	-0.296	—	+0.004	+0.065
$CuSe_2P(OCH_3)_2]_2$	+0.367	+0.130	—	-0.097	-0.290	—	+0.011	+0.073
$AgS_2P(SCH_3)_2$	+0.320	+0.078	-0.112	—	—	—	-0.038	+0.022
$Ni[S_2P(SCH_3)_2]_2$	+0.517	+0.083	-0.108	—	—	—	-0.033	+0.027
$Cu[S_2P(SCH_3)_2]_2$	+0.323	+0.091	-0.101	—	—	—	-0.026	+0.034
$AgSe_2P(SCH_3)_2$	+0.325	+0.082	-0.108	-0.139	—	—	-0.034	+0.026
$Ni[Se_2P(SCH_3)_2]_2$	+0.523	+0.088	-0.104	-0.134	—	—	-0.029	+0.031
$Cu[Se_2P(SCH_3)_2]_2$	+0.328	+0.095	-0.097	-0.128	—	—	-0.022	+0.039

are listed in Table 7. For the ligand anions the calculations were made for the silver(I) salts which were chosen because the calculated charge distributions reasonably approximate those resulting from extended-Hückel molecular orbital calculations [289]. Charge distributions for nickel(II) and copper(II) complexes are also tabulated along with data for comparable complexes and several, at present, hypothetical compounds. The low values of the positive fractional charges for the metal ions are consistent with the low nephelauxetic parameters obtained in the electronic spectra of complexes with sulfur and selenium donors (Section V. 1). The smaller values of the positive charges

Table 7 (continued)

Compound	Metal ion	P	S	Se	O	F	C	H
AgS_2PF_2	+0.461	+0.202	-0.001	—	—	-0.329	—	—
$Ni[S_2PF_2]_2$	+0.700	+0.228	+0.023	—	—	-0.309	—	—
$Cu[S_2PF_2]_2$	+0.503	+0.250	+0.043	—	—	-0.293	—	—
$AgSe_2PF_2$	+0.475	+0.214	—	-0.023	—	-0.320	—	—
$Ni[Se_2PF_2]_2$	+0.717	+0.242	—	+0.002	—	-0.299	—	—
$Cu[Se_2PF_2]_2$	+0.518	+0.264	—	+0.021	—	-0.282	—	—
$AgS_2P(CH_3)_2$	+0.298	+0.058	-0.130	—	—	—	-0.056	+0.003
$Ni[S_2P(CH_3)_2]_2$	+0.492	+0.063	-0.125	—	—	—	-0.052	+0.008
$Cu[S_2P(CH_3)_2]_2$	+0.302	+0.072	-0.117	—	—	—	-0.043	+0.016
$AgSe_2P(CH_3)_2$	+0.304	+0.064	—	-0.156	—	—	-0.051	+0.008
$Ni[Se_2P(CH_3)_2]_2$	+0.499	+0.069	—	-0.151	—	—	-0.046	+0.013
$Cu_2[Se_2P(CH_3)_2]_2$	+0.308	+0.078	—	-0.143	—	—	-0.038	+0.022

for the copper(II) ion than for nickel(II) also reflect the greater ease with which the copper(II) complexes may be converted to copper(I) complexes. It is fairly well-established that copper(I) complexes tend to be predominant with sulfur and selenium ligands; however, the dithio- and diselenocarbamates of copper(II) are fairly stable. This may be due to the greater extent of electron delocalization in those ligands than in the phosphate derivatives. Cavell and his co-workers have found [284] that the complexes $Ni[S_2PX_2]_2$ ($X = OC_2H_5, F, CH_3, C_6H_5$ and CF_3) fall in the spectrochemical order $OC_2H_5 \sim F > CH_3 \sim C_6H_5 \sim CF_3$ and that the metal-sulfur stretching frequencies vary in the order $OC_2H_5 > F > CH_3 > C_6H_5 > CF_3$. The calculated charges for nickel(II) in the complexes (Table 7), which might be considered to reflect the extent of metal-ligand interaction, vary in the order $F > OCH_3 > CH_3$ in reasonably approximate agreement with the available [284] experimental data. Since subtle changes in phosphorus hybridization and coupling of vibrational motions cannot be incorporated into the electronegativity model for the charge distributions, the predictions based on the model can be considered to be in modestly good agreement with laboratory experience. Here we have only wished to show that a relatively simple model can be employed to yield some useful correlations of data for dithiophosphate complexes.

2. Stability

Kabachnik and co-workers [290] have determined the pKa's of a number of dithio- and monothio-organophosphorus acids and considered the thiol-thione tautomeric equilibria of monothiophosphoric acids. For most monothio acids the tautomeric equlibrium is shifted toward the thione form. There are slight changes in the pKa's of O, O'-dialkyldithiophosphoric acids with substituent variation. Other data [114,291] indicates that the pKa values are rather substituent dependent, e.g., Ref. [114] gives pKa values of -1.10 and $+0.22$ for $HSP(S)(OC_2H_5)_2$ and $HSP(S)(On\text{-}C_4H_9)_2$, respectively. Few reliable studies of pKa values of R-dtp acids have been reported. Stability constants for a variety of zinc, cadmium, mercury, copper and nickel R-dtp complexes have been described [116,291-294, 321]. Unfortunately, the data available is not always readily comparable with data for complexes with other ligands. For mercury(II) complexes the stability in 40% ethanol was found [291] to decrease in the order

$$(C_2H_5)_2P(S)S^- > (C_2H_5O)_2PS_2^- > (C_2H_5)_2P(S)O^-$$
$$> (C_2H_5O)_2P(S)O^- > (C_2H_5)_2P(S)OCH_3.$$

An investigation [292] of mercury(II) complexes with the acids $(RO)_2PS_2H$, $(RO)PS_2H$ and R_2PS_2H (where R = ethyl, propyl, isopropyl, and n-butyl) showed that with $(RO)_2PS_2H$ mercury forms HgL_2, HgL_3^- and HgL_4^{2-} type complexes, with $(RO)PS_2H$, HgL_3^- type complexes predominantly and with R_2PS_2H only HgL_2 complexes are found in 40% ethanol.

Shetty and Fernando [295] investigated the polarographic behavior of $Ni(ethyl\text{-}dtp)_2$ at the dropping mercury electrode in ethanol and ethanol-water media. The nickel ion was catalytically reduced in the presence of small quantities of ethyl-dtp at more positive potentials than in the absence of ethyl-dtp. In ethanol a single wave that is almost completely controlled by diffusion was obtained whereas in ethanol-water mixtures, in which the water content was less than 40% by volume, two waves were obtained. The first wave is the catalytic wave that arises from the complexed nickel and the second wave is the reduction wave of uncomplexed nickel(II). Polarographic data for Ni(ethyl-dtp)$_2$ and related complexes has also been reported by Cavell's group [284]. The polarographic oxidation of sodium R-dtp salts to disulfides, $(RO)_2P(S)S-S(S)P(OR)_2$, has been reported [213].

Table 1 does not list all of the Zn(R-dtp)$_2$ complexes which have been prepared. A huge number [64,66,70,79,84,85,101-103,188,191,204,206,210,212,296-305] of Zn(R-dtp)$_2$ complexes have been characterized and their thermal stabilities investigated [173,184,190,297-299,301-305]. Zn(R-dtp)$_2$ compounds are thermally degraded to volatile olefins and non-volatile residues and this serves as the basis for gas chromatographic determination of the compounds [304,305]. Several papers describing pyrolyses of Zn(R-dtp)$_2$ complexes have discussed mechanisms for formation of olefins, sulfides, and other products [173,184,190,298,299,304]. Dakternieks and Graddon [185,283] as mentioned earlier, have reported thermodynamic measurements for depolymerization and adduct formation reactions of zinc, cadmium and mercury R-dtp compounds.

The mass spectra of Ni(R-dtp)$_2$ complexes and a number of amine adducts have been reported [112]. The former gave good mass spectra whereas the spectra of the adducts were difficult to obtain due to their much lower vapor pressure. For the adducts the molecular ion peak was not well-defined. The loss of a neutral species, C_2H_4, from Ni(ethyl-dtp)$_2$ was postulated to occur *via* a McLafferty-type rearrangement and reaction schemes were proposed. Mass spectral data for a variety of compounds related to R-dtp acids, esters and complexes [284,306,307] have also been described. To date, the amount of data is insufficient to draw general conclusions regarding the mass spectra of R-dtp complexes.

The stabilities of metal R-dtp complexes have for the most part not been thoroughly documented. This is rather surprising considering the wide variety of applications of the complexes in analytical chemistry [113-115,308-320].

3. Reactions

Mercury(I) R-dtp complexes are unstable [213] with respect to disproportionation:

$$Hg_2^{2+} + 2 S_2P(OR)_2^- \longrightarrow Hg + Hg[S_2P(OR)_2]_2$$

From the grey reaction product the mercury(II) complex can be isolated by acetone extraction.

Copper(II) R-dtp complexes are unstable with respect to the reaction [107,111,121,159,160,309,313,314]:

$$2 \; Cu(R\text{-}dtp)_2 \longrightarrow 2 \; Cu(R\text{-}dtp) + (RO)_2\overset{\displaystyle S}{\overset{\displaystyle \|}{P}}-S-S-\overset{\displaystyle S}{\overset{\displaystyle \|}{P}}(OR)_2$$

however, in some instances, small amounts of the copper(II) chelates can be trapped [159] in the isomorphous nickel(II) chelates by preparing the chelates using a mixed solution of the metal salts, adding the ligand acid or its potassium salt and extracting the mixture of chelates into a non-aqueous solvent. The copper(II) chelates are stable when dissolved in various non-aqueous solvents for periods up to several hours [121]. Copper(II) ion reacts with Ni(ethyl-dtp)$_2$ [309,314] and Zn(isopropyl-dtp)$_2$ [313] to yield colored solutions (copper(II) R-dtp complexes??) which are suitable for the colorimetric determination of copper. The copper(II) complex with phenyl-dtp, reportedly, is also light sensitive [311].

Vanadium(III) ethyl-dtp can be prepared [118,119] by interaction of the acid with a degassed solution of vanadium(III) chloride in absolute methanol. The complex is both water and oxygen sensitive. Reaction of the acids, ethyl-dtp or *n*-propyl-dtp, with vanadyl(IV) sulfate pentahydrate involves reduction to the vanadium(III) *tris*-chelates [118,121]. The reaction of V_2O_5 or $VOCl_3$ with HS_2PF_2 yields [120] $V[S_2PF_2]_3$ with some $VO[S_2PF_2]_2$ impurity.

Amine adducts of Ni(R-dtp)$_2$ complexes have been extensively examined in the solid state [166-170,260] and solution [112,132,143-148,150-158,172,322,323,325,326, 328]. In solution two types of equlibria are of principal concern, *i.e.*,

$$\text{Ni(R-dtp)}_2 + \text{amine} \xrightarrow{K_1} [\text{Ni(R-dtp)}_2 \cdot \text{amine}]$$

$$[\text{Ni(R-dtp)}_2 \cdot \text{amine}] + \text{amine} \xrightarrow{K_2} \text{Ni(R-dtp)}_2(\text{amine})_2.$$

The *bis*-amine adducts characterized are green, paramagnetic, *trans*-octahedral species whereas the nature of the mono-amine adducts has been more controversial [143,147,148,152,158]. Stability constant measurements [144,153,156,158] for the above reactions have shown that $K_1 \ll K_2$ for pyridine and primary amines and that secondary amines and sterically crowded amines, *e.g.*, 2-picoline, form monoadducts with very small association constants. NMR studies [158] of Ni[ethyl-dtp]$_2$ with amines in deuteriochloroform solution indicated that the five-co-ordinate adducts of 2-picoline and 2,6-lutidine are diamagnetic whereas the diethylamine and di-*n*-butylamine monoadducts are paramagnetic. In view of the magnitude of the formation constants it is possible that Carlin and Losee's solutions [158] did not contain enough 2-picoline and 2,6-lutidine to give rise to observable NMR contact shifts. Sacconi [327] correlated the sum of the Allred-Rochow electronegativities, $\Sigma\chi$, for various $X_nY_mZ_{5-n-m}$ and other chromophore groupings for five-coordinate cobalt(II) and nickel(II) complexes with the known magnetic behavior of various complexes. When $\Sigma\chi \cong 12.8$ the transition from high-spin to low-spin magnetic behavior could be expected to occur. This result (see Table 8) suggests that five-co-ordinate nickel(II) complexes with NS$_4$ and OS$_4$ groupings should be high-spin paramagnetic species. NMR data [143,147,148,158,322,323,326] for a variety of adducts as well as magnetic susceptibility measurements [152] for the lemon-yellow solid adduct, Ni[ethyl-dtp]$_2$·piperidine, support the high-spin nature of the five-co-ordinate amine adducts of Ni[R–dtp]$_2$ complexes. Table 8 also indicates the expected spin behavior of five-co-ordinate adducts not yet characterized. The electronic spectra of the five-co-ordinate amine adducts, discussed in Section V. 1., have been interpreted [148] in terms of a square pyramidal geometry with approximate C_{4v} symmetry. Molecular models of many of the adducts indicate that this form would be preferred to the possible trigonal bipyramidal structure. Furlani [330] has noted that a high-spin bipyramidal geometry with sulfur ligands would be unusual.

Table 8. Allred-Rochow electronegativity and magnetic behavior of NiL$_4$X complexes[a]

Group	ΣX	Susceptibility	Group	ΣX	Susceptibility
S$_4$O	13.26	High	As$_4$O	12.30	–
S$_4$N	12.83	High	As$_4$Cl[b]	11.63	Low
S$_4$S	12.20	–	As$_4$Br[b]	11.54	Low
S$_4$As	11.96	–	As$_4$I[b]	11.01	Low
S$_4$P	11.82	–	As$_5$[b]	11.00	Low

Crossover from high to low-spin: $\Sigma X \cong 12.8$.

[a] Sacconi, L.: J. Chem. Soc. (A) *1970*, 248.

[b] Preer, J. P., Gray, H. B.: J. Am. Chem. Soc. *92*, 7306 (1970).

In our NMR studies [143,147,148,322–324] of amine and other adducts of Ni[R–dtp]$_2$ complexes neat amines were employed in order to eliminate variations in extent of association (H-bonding) of the amines, to permit observation of NH proton shifts, and to maximize the concentration of the preferred adduct. The use of high concentration of primary amines in solutions with Ni[R–dtp]$_2$ complexes can lead to products other than those expected, *e.g.*, with aliphatic diamines, the R-dtp anion salts of *tris*(diamine)nickel(II) chelates are obtained [145]. Furlani and co-workers [154] have shown that Ni-(ethyl-dtp)$_2$ reacts with *n*-butyl amine to yield complexes containing the NiS$_2$N$_4$ chromophore, presumably with monodentate ethyl-dtp. In all work with adducts it is necessary to assure that the complexes, adduct molecules and solvent systems are anhydrous. A number of authors [132,284,295,329] have shown that Ni[R–dtp]$_2$ complexes decompose when in contact with water.

A number of thermodynamic studies of Ni[R–dtp]$_2$ adducts with amines have been described [112,143,156,158] Dakternieks and Graddon [156] have measured the enthalpies of addition of pyridine, 4-picoline, 2,2'-dipyridyl and 2,9-dimethyl-1,10-phenanthroline to Ni[ethyl-dtp]$_2$ and pyridine to Ni[propyl-dtp]$_2$. The sums of the stepwise enthalpies for pyridine addition to the ethyl and propyl complex differ by $4kJ$/mole, the total enthalpy being larger for the formation of the *bis*-pyridine adduct of the ethyl derivative. Thermogravimetric derterminations [112)] of the *bis*-pyridine adducts of $Ni[S_2P(OR)_2]_2$ (R = CH_3, C_2H_5 and C_6H_{11}) indicate there is a stepwise loss of pyridine with considerable overlap of peaks and that the five-co-ordinate mono-pyridine adducts lose pyridine rapidly. These results are in accord with expectations from attempts to synthesize stable five-co-ordinate adducts. The energies associated with the reaction:

$$Ni[\text{isopropyl-dtp}]_2 \cdot 2L \longrightarrow Ni[\text{isopropyl-dtp}]_2 + 2L_{(g)}^{\uparrow}$$

were found to be 24.0, 25.3 and 15.4 kcal/mole for L = pyridine, 3-picoline, and 4-picoline, respectively [143)]. The values for the pyridine and β-picoline adducts are comparable with values obtained from related NiL_4Cl_2 by Beech and co-workers [331)]. Some preliminary differential scanning calorimetric data [150)] has been obtained for green *cis*-Ni[R-dtp]$_2$(2,2'-bipyridyl) complexes (Fig.2).

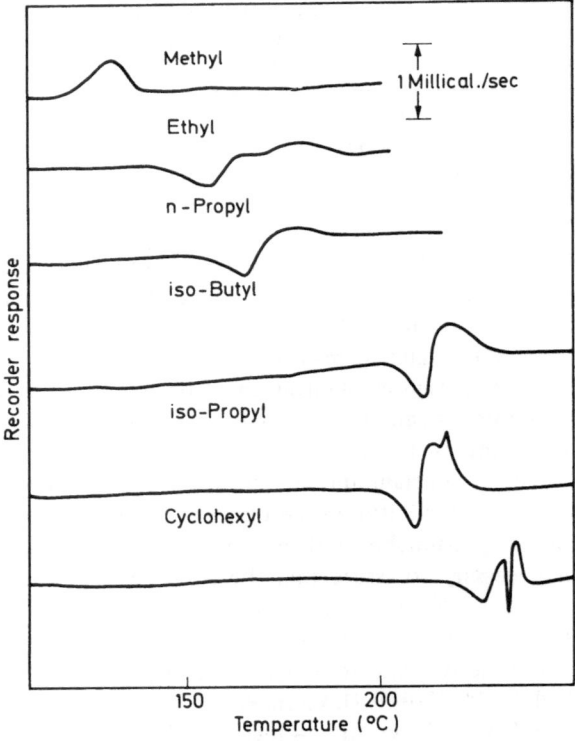

Fig. 2. Differential scanning calorimetry of green *cis*-Ni[S$_2$P(OR)$_2$]$_2$-2,2'-dipyridyl) complexes. The R-groups are indicated

The results show that as the R substituent increases in size it becomes possible to see a reaction pattern in which there is an endothermic reaction followed by an exothermic reaction followed by an endothermic reaction which in turn is followed by an exothermic reaction. This reaction sequence is particularly noticeable for the cyclohexyl derivative and is tentatively interpreted by the following scheme:

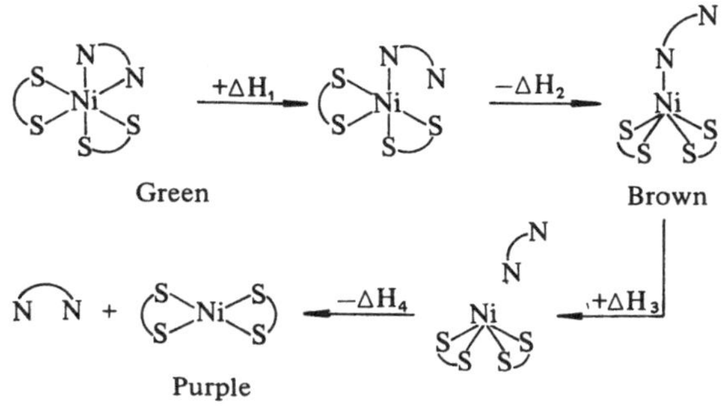

The first and third reactions involve a breaking of Ni-N bonds whereas the second step involves an exothermic rearrangement to a square pyramidal species and the final step an exothermic rearrangement to the diamagnetic, purple parent complex. The color changes accompanying the dsc runs are indicated above.

Ni[R-dtp]$_2$ complexes form paramagnetic five-co-ordinate adducts with hexamethylphosphoramide [324]. To date no phosphate esters have been found to form adducts. However, the NMR spectra of solutions [145] of Ni(R-dtp)$_2$ complexes after long standing or gentle heating clearly show that scrambling reactions of the type:

$$\text{Ni[R-dtp]}_2 + \text{O} = \text{P(OR')}_3 \rightleftharpoons \text{Ni[R'-dtp]}_2 + \text{O} = \text{P(OR')}_{3-n}\text{(OR)}_n$$

occur. The lability of the OR groups in Ni(R-dtp)$_2$ complexes is also evident from the reaction [145]:

$$\text{Ni[cyclohexyl-dtp]}_2 + \text{piperidine} \xrightarrow[\text{1 h}]{\text{reflux}}$$

$$\text{Ni(cyclohexyl-dtp) (S}_2\text{P\{cyclohexyl\}\{piperidyl\})}$$

$$\xrightarrow{\text{reflux 3 days}} \text{Ni[S}_2\text{P(cyclohexyl) (piperidyl)]}_2$$

where cyclohexyl groups have been replaced by piperidyl groups. Dakternieks and Graddon [185] obtained Cd[S$_2$P(NH-cyclohexyl)$_2$]$_2$·4H$_2$O upon refluxing Cd[ethyl-dtp]$_2$ with cyclohexylamine for 40 min. Reactions of this type bear further study.

Finally, it is noted that Au(III) is also reduced by R-dtp ligands. Yellow crystalline $\{[AuS_2P(O\text{-isopropyl})_2]_2\}_n$ is prepared [175] by the reaction of an aqueous solution of the ammonium salt of isopropyl-dtp with an aqueous solution of $HAuCl_4$.

V. Spectroscopic Properties

1. Electronic Spectra

Jørgensen [6] has written an authoritative review of the electronic spectra of sulfur-containing ligands and his article provides a wealth of comparative data. As Jørgensen [6] has noted, R-dtp ligands do not absorb as near to the visible region as other sulfur donor ligands. The ethyl-dtp anion has its first absorption band at 44.5kK ($1k$K = 1000 cm^{-1}). This makes R-dtp complexes particularly suitable for studies of electronic charge transfer transitions.

Reports dealing with the electronic spectra of R-dtp complexes are listed in Table 9. Several papers [120,122,123,252,284] have been concerned with the place of R-dtp in the spectrochemical and nephelauxetic series, particularly with reference to other sulfur ligands. The azide ion has been classified [340] in the spectrochemical series between S-bonded thiocyanate and ethyl-dtp and in the nephelauxetic series between bromide and ethyl-dtp. The spectrochemical order of the ligand $-S_2PX_2$ according to X has been found [341] to be

$$F \sim OC_2H_5 > CH_3 \sim C_6H_5 \sim CF_3.$$

The nephelauxetic series of $-S_2PX_2$ ligands according to X in order of B; is

$$F \sim OC_2H_5 < CH_3 \sim C_6H_5 \sim CF_3$$

indicating greater electronic delocalization with those substituents found to produce the largest phosphorus electron spin-nuclear spin hyperfine coupling constants in the electron spin resonance spectra of the vanadyl(IV) complexes [252].

Cavell and his co-workers have reported [252] the electronic spectra of

$$OV(S_2PX_2)_2 \quad (X = CH_3, C_6H_5, OC_2H_5, F \text{ and } CF_3)$$

complexes. The optical bands were resolved using Gaussian analyses and the oscillator strengths evaluated. If Band II is assigned to 10Dq, as is usually done [343], the spectrochemical series formed by these ligands in order of the substituent X is

$$CF_3 < CH_3 \sim C_6H_5 < OC_2H_5 < F.$$

This ordering is valid only if the angle, θ, between O = V and the SVS bi-

Table 9. Electronic spectra of metal R–dtp complexes

Ion	Ref.[a]
Ligand anions	202,332)
VO(IV)	252)
V(III)	118[a],119[a],120,132,341)
Zr(IV)	127)
Nb(IV)	127)
Nb(V)	126)
Cr(III)	122,123,125,132,234[a]),337[a]),
	338,339,341,342)
Mo(IV)	130)
Mo(V)	131)
Fe(III)	6,132,135,136)
Co(II)	132,138,223,341)
Co(III)	132,138,234[a]),334,335,341)
Rh(III)	139,234[a])
Ir(III)	139,234[a])
Ni(II)	112,132,147,149[a]),157,284,
	333,336,341))
Ni(II)-adducts	112,132,143,144,147,148,
	152-154,158,323-325)
Pd(II)	284,310,341)
Pt(II)	284,341)
Cu(II)	311)

[a] Single-crystal data.

sector is constant for all of the complexes. The energy of Band II of vanadyl(IV) complexes can be readily shown [344] to be expressed by:

$$\nu_2 = \frac{10}{6} \, \alpha_4^E \, \sin^4 \theta$$

where α_4^E (= 6 Dq) is the crystal field splitting parameter for the equatorial ligands. Unless θ is identical for a series of vanadyl(IV) complexes with approximate C_{4v} symmetry, the spectrochemical series obtained is highly tenuous.

Vanadium(III) complexes have been the subject of several recent studies [118 -120,341]. The electronic spectra of these complexes in methylene chloride exhibit the two expected $^3T_{2g}(F) \leftarrow {}^3T_{1g}(F)$ and $^3T_{1g}(P) \leftarrow {}^3T_{1g}(F)$ transitions as well as the two-electron $^3A_{2g}(F) \leftarrow {}^3T_{1g}(F)$ transition which is forbidden in octahedral symmetry and generally not observed in solution spectra. The presence of this latter band is indicative of the presence of a nonoctahedral crystal field which is expected for these complexes since the crystal structure

[118] of V(ethyl-dtp)$_3$ has shown the V(III) ion to be in a field of D_3 symmetry. The spectrochemical series for a number of ligands with V(III) was found [120] to be

$$-S_2P(CH_3)_2 \quad \sim \quad -S_2P(C_6H_5)_2 \quad \sim \quad -S_2P(CH_3)_2 \quad < \quad -S_2PF_2 \quad < \quad -S_2P(OC_2H_5)_2.$$

The nephelauxetic series was found [120] to be

$$-S_2P(C_6H_5)_2 \quad < \quad -S_2P(CF_3)_2 \quad < \quad -S_2P(CH_3)_2 \quad \sim \quad -S_2P(OC_2H_5)_2 \quad < \quad -S_2PF_2$$

although the variations are hardly significant.

The intensities of polarized ligand field spectra [118,119] of V(ethyl-dtp)$_3$ doped into the corresponding indium(III) compound exhibit a relatively small temperature dependence. The source of the large intensities of the "d–d" transition is the static distortion of the ligand field [119] and not vibronic effects to any appreciable extent.

The electronic spectra of niobium(IV) and -(V) and zirconium(IV) complexes [126,127] have been reported but not interpreted. The spectrum of Nb(ethyl-dtp)$_4$ is of particular interest since the compound is probably 8-coordinate. Discussion of the spectrum of binuclear molybdenum complexes [130,131] employed the molecular orbital model of Blake, Cotton and Wood for $MO_2O_3L_x$ complexes [345].

The optical spectra of Cr(R-dtp)$_3$ complexes have been extensively investigated. A typical spectrum is shown in Fig. 3. The electronic spectra of

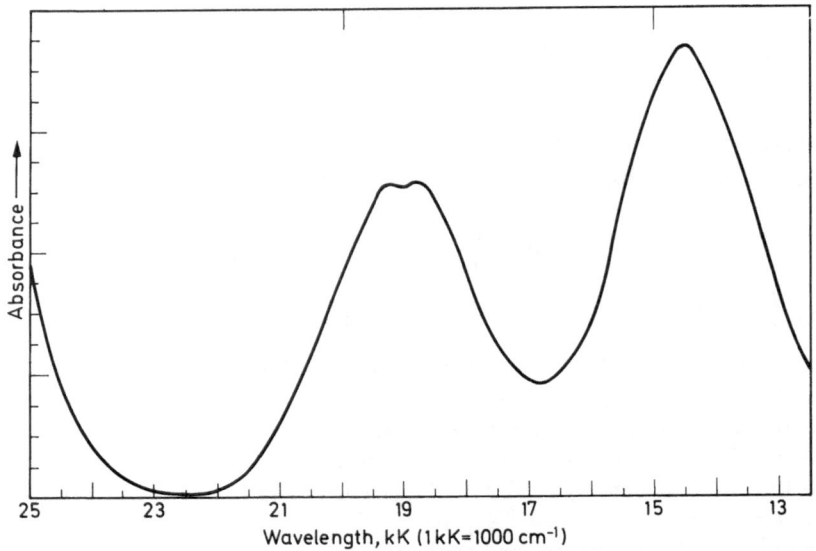

Fig. 3. Electronic spectrum of Cr(methyl-dtp)$_3$ in chloroform. The band at about 19kK is split by 0.4kK

$Cr(R\text{-}dtp)_3$ complexes exhibit two strong absorption bands, on at $\sim 14.5\ kK$ ($1kK = 1000\ cm^{-1}$), assigned to the $^4T_{2g} \leftarrow {}^4A_{2g}$ transition, and one at $\sim 19.1kK$, assigned to the $^4T_{1g}(F) \leftarrow {}^4A_{2g}$ transition in octahedral symmetry. From the data in Table 10, in Ref. [122], and Cavell's work [123] the following spectrochemical series is obtained for chromium(III):

$$^-S_2As(CH_3)_2 \quad < \quad {}^-S_2P(C_6H_5)_2 \quad < \quad {}^-S_2P(CH_3)_2 \ = \ {}^-S_2P(CF_3)_2 \quad < \quad S \ =$$

$$= \ C \underset{\underset{H}{\diagdown}N-CH_3}{\overset{\overset{H}{\diagup}N-CH_3}{\diagup}} \quad < \quad {}^{2-}S \quad < \quad {}^-S_2PF_2 \quad \sim \quad R-dtp \quad < \quad S =$$

$$= \ C \underset{\diagdown NH_2}{\overset{\diagup NCH_3}{\overset{H}{}}} \quad < \quad {}^{2-}S_2C_2(CN)_2 \ = \ {}^-S(Se)CN(C_2H_5)_2 \quad < \quad {}^-S_2CN(C_2H_5)_2 =$$

$$= \ {}^-S_2CNH_2 \quad < \quad {}^-S_2COC_2H_5 \quad < \quad {}^{2-}S_2C_2O_2.$$

In Fig. 3 it is noted that the band at $\sim 19.1kK$ is split some $0.4kK$. This was originally thought to arise from a trigonal ligand field component [122]. However, examination of single crystal polarized spectra [119,132,234,338,339] resulted in the conclusion that the splitting arises from the presence of spin-forbidden transitions whose intensity is enhanced because of proximity to spin-allowed ones, and not trigonal splitting. Magnetic circular dichroism (MCD) measurements [338] have shown that the trigonal splitting of the $^4T_{2g}$ level is ca. 0.5 kK. Schreiner and his co-workers [338] have discussed the difficulties in assigning the energies of the higher excited states of $Cr(III)S_6$ chromophores. Emission from $Cr(ethyl\text{-}dtp)_3$ could not be obtained [339] at 83 °K. Cancellieri *et al.* [125] in their study of luminescence spectra of $Cr(III)S_6$ complexes concluded that the vertical energy difference $E(T_{2g}) - E(^2E_g)$ ranges from $\sim 3kK$, when only phosphorescence is observed, to $< 1kK$, where only fluorescence occurs. In the intermediate case, *e.g.* $Cr[S_2CN(C_2H_5)_2]_3$, simultaneous phosphorescent and fluorescent emission is observed.

Table 10 also collects intensity (oscillator strength) data for some $Cr(III)S_6$ complexes. Generally, only absorption maxima and molar extinction coefficients are reported for metal complexes and organic compounds and extinction coefficients are frequently employed as a measure of intensity. However, as is well-known [346], the oscillator strength is a more fundamental measure of band intensity since it can be related to transition moment integrals. The oscillator strengths of absorption bands are readily obtained and it is most unfortunate that most authors neglect to include such information in their publications. Admittedly, the oscillator strength data for many complexes will be a composite value since several electronic transitions occur under a given band

J. R. Wasson, G. M. Woltermann and H. J. Stoklosa

Table 10. Visible spectra of Cr(III)S$_6$ complexes

Compound	$\nu_1(kK)$	ϵm_1 [a]	$10^5 f_1$ [b]	$\nu_2(kK)$	ϵm_2 [a]	$10^5 f_2$ [b]
Cr(methyl-dtp)$_3$ [c]	14.45	372	474	19.05	255	360
Cr(isopropyl-dtp)$_3$ [c]	14.47	348	454	19.05	250	353
Cr(isobutyl-dtp)$_3$ [c]	14.45	358	460	19.05	256	359
Cr[S$_2$As(CH$_3$)$_2$]$_3$ [d]	13.47	285	332	18.20	196	256
Cr[SSeCN(C$_2$H$_5$)$_2$]$_3$ [e]	15.15	209	372	19.61	246	294
NaCrS$_2$ [f]	14.20	–	–	18.87	–	–

[a] Molar absorptivity.
[b] Oscillator strengths were calculated using the expression $f = 4.60 \times 10^{-9}\, \epsilon_{max}\nu_{1/2}$ where ϵ_{max} is the molar absorptivity of the band maximum and $\nu_{1/2}$ is the band width at half-height expressed in wavenumbers [346].
[c] Ref. [122].
[d] Kuchen, W., personal communication.
[e] Kirmse, R., personal communication.
[f] Companion, A. L., Mackin, M.: J. Chem. Phys. *42*, 4219 (1965).

envelope. Nevertheless, such information is valuable when comparing similar complexes. Two chemically significant types of information are to be found in oscillator strength data:

a) Covalency — the more covalent the bonding in a complex, the greater the expected oscillator strength.

b) Symmetry — the more the geometry deviates from the centro-symmetric case, the greater the expected oscillator strength.

Of course these generalizations must be used with caution; however, they are useful in comparing similar compounds. For a series of octahedral Cr(III) complexes it has been found that the oscillator strength, f, varies approximately linearly with the nephelauxetic parameter, β, also a measure of covalency [342].

The nephelauxetic parameter, β, for Cr(III) complexes can be evaluated from the first two visible absorption bands using the expression:

$$\beta = \frac{B}{918 \text{ cm}^{-1}}$$

where
$$B = \frac{2\nu_1^2 - 3\nu_1\nu_2 + \nu_2^2}{15\nu_2 - 27\nu_1}$$

where ν_1 and ν_2 are the energies (cm^{-1}) of the $^4T_{2g} \leftarrow {}^4A_{2g}$ and $^4T_{1g}(F) \leftarrow {}^4A_{2g}$ transitions, respectively. Besides being able to correlate β with oscillator strength, f, of transitions [342], the β values may be employed [342] to evaluate the effective charges on the Cr(III) ions in various complexes. Fig. 4 provides plots of the effective metal charge *vs.* β, the nephelauxetic parameter and B, the Racah interelectronic repulsion parameter for chromium(III). Jørgensen

[347] showed that for the $3d$ transition metals the variation of the Racah inter-electronic repulsion parameter, B, with cationic charge, Z, and q, the occupation number of the d^q shell, is well-expressed by the relation:

$$B(\text{cm}^{-1}) = 384 + 58q + 124(Z+1) - 540/(Z+1)$$

Using this expression and β values for Cr(III) complexes it has been shown [342] that CrO$_6$, CrN$_6$ and CrS$_6$ chromophores have effective metal charges in the ranges 0.8 – 1.4, 1.19 – 2.14 and 0.56 – 1.12, respectively.

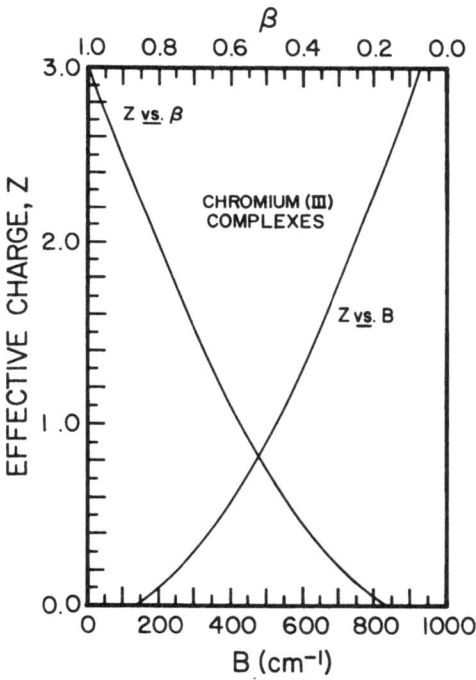

Fig. 4. Effective metal charge, Z, in chromium(III) complexes as a function of the nephelauxetic parameter, β, and the Racah interelectronic repulsion parameter, B

    Iron(III) *tris*-methyl-dtp, prepared from anhydrous iron(III) chloride and the ammonium salt of the ligand in absolute methanol, decomposes quickly [6,132]. The complex is purple in high dilution in In(methyl-dtp)$_3$ and has charge transfer bands in the region 17–19kK. Limited spectroscopic data is available for Fe(R-dtp)$_3$ complexes although a number of compounds [132,134, 136] have been reported. The blue-green complex *trans*-Fe(isopropyl-dtp)$_2$(pyridine)$_2$ has been prepared [136] but spectroscopic data for this compound is not available. Raspberry red Ru(ethyl-dtp)$_3$, a low-spin $4d^5$ system [137], absorbs at 19.55, 23.70, 25.00, 35.70 and 41.70kK. The first three transitions are ascribed to "$d{\to}d$" transitions.

Cobalt(II) R-dtp complexes have been the subject of several studies [138, 223,341]. Co(ethyl-dtp)$_2$ occurs in a tetrahedral non-solvated form in carbon tetrachloride [138] but undergoes solvation in other non-aqueous solvents [138, 341]. The spectrochemical series for tetrahedral Co(S$_2$PX$_2$)$_2$ (where X = F, OC$_2$H$_5$, CF$_3$, CH$_3$, C$_6$H$_5$) was found [341] to be

$$F \sim OC_2H_5 > CF_3 \sim CH_3 \sim C_6H_5 .$$

The spectrochemical series

$$^-S_2P(OC_2H_5)_2 > ^-SOP(OC_2H_5)_2 > ^-SeOP(OC_2H_5)_2$$

was found by Larionov and Il'ina [223].

The cobalt(II) complexes with R-dtp and related ligands are readily oxidized by air and other oxidizing agents to yield the corresponding cobalt(III) *tris*-chelates. Electronic spectra and ^{59}Co NMR spectra data [334,335] have shown that the spectrochemical series of sulfur ligands is:

$$^-S_2COC_2H_5 > ^-S_2CSC_2H_5 > ^-S_2CN(C_2H_5)_2 > ^-S_2P(OC_2H_5)_2 .$$

Cavell and co-workers [341] have established the spectrochemical series: $^-S_2PF_2 \sim {}^-S_2P(OC_2H_5)_2 > {}^-S_2P(CH_3)_2 \sim {}^-S_2P(C_6H_5)_2 \sim {}^-S_2P(CF_3)_2$ for cobalt(III) complexes. Lebedda and Palmer [119] and Tomlinson [234] have studied the single-crystal polarized spectra of Co(ethyl-dtp)$_3$ doped into the indium(III) complex. The intensity of the transitions comes from the distortions of the octahedral ligand field and is not due to appreciable vibronic effects. Hillis and DeArmond [139] have measured the low temperature emission and absorption spectra of Rh(ethyl-dtp)$_3$ and Ir(ethyl-dtp)$_3$. The emission for all complexes is broad, structureless and occurs in the near-infrared. Emission lifetimes are all of $\sim \mu$sec magnitude. The emission is assigned [139] as a metal localized $^3T_1 \to {}^1A_1$ transition. Jørgensen [132] has also characterized the absorption spectra of Rh(III) and Ir(III) chelates with ethyl-dtp.

The most extensively examined absorption spectra have undoubtedly been those of the Ni(R-dtp)$_2$ complexes and their adducts. Subsequent to Jørgensen's original study of Ni(ethyl-dtp)$_2$, single-crystal studies have been described [149,159,157] which disagree with regard to the assigned energy levels. Molecular orbital calculations (not available to the reviewers) of the energy levels of Ni(ethyl-dtp)$_2$ and related ethylxanthate and diethyldithiocarbamate complexes have also been described [336]. The spectra of purple Ni(R-dtp)$_2$ complexes show two fairly strong bands [112,132], one at 14.65-15.60 kK, and the other at 19.05 – 19.35 kK with a further absorption which is sometimes discernible as a shoulder at $\sim 25k$K on the charge transfer band in the ultraviolet region.

The charge transfer band has been assigned to the $3p \rightarrow 4s$ excitation of sulfur [132]. Assuming D_{2h} effective symmetry for Ni(cyclohexyl-dtp)$_2$ the assignments $^1B_{1g} \leftarrow {}^1A_g$ and $^1B_{3g} \leftarrow {}^1A_g$ were made [146,147] for the bands at 14.50 and 19.20 kK, respectively, in analogy with the assignments [3,348] for Ni(S$_2$C$_2$(CN)$_2$)$_2^{2-}$. These assignments are confirmed by the single-crystal studies [149,157] although some of the other band assignments are still somewhat controversial. Müller and his group [349] have prepared salts containing the bis(tetrathiotungstato)nickel(II) anion, Ni(WS$_4$)$_2^{2-}$, which exhibits bands at $\sim$ 14.3 and $\sim$ 19.0kK. the spectrochemical series for NiS$_4$ complexes is [350]:

$$S_2C_2O_2^{2-} > S_2C_2CH_2O_2^{2-} > \text{2,3-dimercaptopropanol anion}$$
$$> {}^-S_2CN(C_2H_5)_2 > {}^-S_2COC_2H_5 > \text{dithioacetylacetonate}$$
$$> \text{R-dtp} > {}^{2-}S_2C_2(CN)_2.$$

The energy levels of NiS$_4$ complexes have also been discussed by Dingle [351] in his analysis of crystal spectra of bis(diethyldithiocarbamato)nickel(II), Ni(dtc)$_2$. The spectrum of the selenium analogue of Ni(dtc)$_2$ has also been reported [352]. The MCD spectrum of Ni(ethyl-dtp)$_2$ has been described [333] and interpreted in terms of D_{4h}, rather than D_{2h}, symmetry. Theoretical studies of low-spin planar d^8 complexes have also been reported [353]. As mentioned earlier, the complex Ni[SP(CH$_3$)$_2$N(CH$_3$)$_2$PS]$_2$ presents an interesting example of a tetrahedral NiS$_4$ chromophore [46-48]. Assignments of the palladium(II) and platinum(II) R-dtp complexes [132,284] have not been as exhaustively studied as those of the nickel(II) compounds.

Spectra of adducts of Ni(R-dtp)$_2$ complexes have been extensively examined [112,138,143,144,147,148,151,152,154,156,158,323-326]. The spectra of the green paramagnetic (cis and trans) adducts with mono and bidentate amines have been interpreted in terms of the octahedral model rather than in terms of their correct molecular symmetries. Some data for amine adducts is given in Table 11. Calculations of the optical spectra of the five-co-ordinate amine adducts have not been in agreement [148,151]. Some of the amine adducts are green trans-octrahedral and exhibit [143] tetragonal splitting of about 0.9kK. The five co-ordinate adducts exhibit two well-defined bands at about 13kK and $\sim$ 21kK with oscillator strengths (see Table 11) which are distinctly greater than those of any of the "$d \leftarrow d$" transitions of the octahedral amine adducts. The larger oscillator strengths are consistent with the lower symmetry ($\sim C_{4v}$) of the five-co-ordinate adducts and in most circumstances can be employed to distinguish between the types of adducts possible. Additional work with the five-co-ordinate adducts will be necessary before band assignments can be considered settled.

The spectra of copper(II) complexes with R-dtp ligands remain an unsettled problem in view of the tendency of Cu(R-dtp)$_2$ complexes to disproportionate. The recent discussion [354] of the energy levels of bis(diethyldithiocarbamato)copper(II) may serve as a useful starting point for discussions of the spectra of Cu(R-dtp)$_2$ compounds.

Table 11. Electronic spectra data [a]

Compound	Solvent	ν_1, kK	em_1	$10^6 f_1$	ν_2, kK	em_2	$10^6 f_2$	ν_3, kK	em_3	$10^6 f_3$
		Octahedral adducts								
Ni(isopropyl-dtp)$_2$ [b]	Pyridine	8.55	14	115	9.01	14	186	14.93	15	183
Ni(isopropyl-dtp)$_2$ [b]	β-Picoline	8.55	13	121	9.01	14	196	14.93	14	214
Ni(isopropyl-dtp)$_2$ [b]	γ-Picoline	8.55	12	111	9.01	13	185	14.93	16	262
Ni(cyclohexyl-dtp)$_2$ [c]	n-Propylamine	9.66	6	99	–	–	–	16.50	20	229
		Five co-ordinate adducts								
Ni(isopropyl-dtp)$_2$ [b]	α-Picoline	12.82	59	798	–	–	–	21.83	238	2848
Ni(cyclohexyl-dtp)$_2$ [c]	Morpholine	14.10	36	583	–	–	–	22.60	85	1480
Ni(cyclohexyl-dtp)$_2$ [c]	Pyrrolidine	15.00	23	367	–	–	–	27.10	96	1330
Ni(cyclohexyl-dtp)$_2$ [c]	Piperidine	14.70	30	906	–	–	–	22.20	63	1220

a) For definitions see Table 10.
b) Ref. 143).
c) Ref. 147).

With diselenophosphates (see Table 4) the ligand field strengths tend to be smaller and the nephelauxetic effect more pronounced. The crystal spectra of Cr(ethyl-dsep)$_3$ has been interpreted [234] in terms of D_3 site symmetry for the metal ion. The electronic spectra of a variety of ethyl-dsep complexes have been reported [231-234]. Octrahedral high-spin Ni(ethyl-dsep)$_2$(pyridine)$_2$ has [232] absorption bands at 8.7, 8.9 and 13.9kK. The spectrochemical series for rhodium(III) was found to be [232, 233]

$$ethyl - dsep \; < \; ethyl - dtp \; < \; {}^-Se_2C - N(C_2H_5)_2.$$

The reducing character of the ligand is linearly related to the relative position of the first spin-allowed transition and the first electron transfer transition. Jørgensen [233] calculated the optical electronegativity of ethyl-dsep as 2.6 which is close to that of ethyl-dtp, 2.7. Jørgensen [233] has also discussed electron delocalization in $M(X_2P)_3$ (X = S, Se) chromophores in terms of a molecular orbital model.

2. Vibrational Spectra

Corbridge [358] has given an extensive review of the infrared spectra of phosphorus compounds. A review exclusively concerned with the vibrational spectra of organophosphorus compounds has also appeared [359]. Many studies containing infrared spectra of R-dtp complexes have been reported [112,118,122,123, 126,127,129-131,133,140,141,146,156,171,182-185,189,191,193,284,300,341,360-367], particularly with respect to the identification of the zinc complexes added to petroleum products. Virtually all of the band assignments made have been empirical since no normal coordinate analyses of any of the complexes have been reported. Additional x-ray structural studies coupled with normal coordinate analyses of the infrared and Raman spectra of the complexes are needed before the proposed band assignments can be taken other than *cum grano salis*.

P$\doteq$S stretching modes are expected to appear in the region 500 – 750 cm^{-1}. This wide range is due to the partial double bond character of the P–S bonds in the complexes. Normally, P=S stretching frequencies are expected in the range 750 – 550 while P–S single bond frequencies can be found in the range 400 – 620 cm^{-1}. P–O–C linkages have a number of vibrations covering a range of about 1250 – 700 cm^{-1}. These frequencies are fairly dependent on the substituents on carbon. Bands found in the region 200 – 400 cm^{-1} are expected to be due to M–S stretches. It has been suggested [360] that M–S stretching frequencies are mainly dependent on the metal. It is noted that the M–S stretches which have been assigned to R-dtp complexes differ slightly from those of other complexes [368-370] allowing relative strengths of metal-ligand bonds to be assessed. Livingstone [112], Adam [360] and Cavell and his co-workers [120,123,283, 284,341], Husebye [317] and Rockett [189] have described studies of particular value in the assignment of vibrational modes of R-dtp complexes. Cavell and his co-workers [284] found a linear correlation between the square of the Ni–S stretching frequency and the first d–d band in the electronic spectrum for a number of $Ni(S_2PX_2)_2$ (X = CF_3, C_6H_5, CH_3, F, OC_2H_5) complexes. This sort of

Table 12. P–S stretching frequencies (cm^{-1})

Compound	$\nu(P-S)$ cm^{-1}	Ref.
Ni(ethyl-dtp)$_2$	644, 545	[360]
Pd(ethyl-dtp)$_2$	630, 536	[360]
Hg(ethyl-dtp)2	658, 567	[360]
Pb(ethyl-dtp)$_2$	662, 571	[360]
Cr(ethyl-dtp)$_3$	656, 548	[360]
Rh(ethyl-dtp)$_3$	645, 546	[360]
In(ethyl-dtp)$_3$	652, 532	[360]
Bi(ethyl-dtp)$_3$	620	[360]
Ni(methyl-dtp)$_2$	648, 524	[112]
Ni(cyclohexyl-dtp)$_2$	640, 527	[112]
Mn(CO)$_4$(phenyl-dtp)	657, 644, 535 (broad, medium peaks in Nujol mull)	[203]
Ni[S$_2$PS$_2$C$_2$(C$_6$H$_5$)$_2$]$_2$	571, 517, 498, 467, 445	[171]

Table 13. Metal-sulfur stretching frequencies (cm^{-1})

Complex	$\nu(M-S)$ cm^{-1}	Ref.
Ni[S$_2$PS$_2$C$_2$(C$_6$H$_5$)$_2$]$_2$	349, 323	[171]
Ni(ethyl-dtp)$_2$	327, 226	[360]
Pd(ethyl-dtp)$_2$	312, 221	[360]
Hg(ethyl-dtp)$_2$	280, 239	[360]
Pb(ethyl-dtp)$_2$	291, 237	[360]
Cr(ethyl-dtp)$_3$	313	[122,360]
Rh(ethyl-dtp)$_3$	293	[360]
In(ethyl-dtp)$_3$	286	[360]
Bi(ethyl-dtp)$_3$	271	[360]
Ni(methyl-dtp)$_2$	355, 325	[112]
Ni(cyclohexyl-dtp)$_2$	359, 338	[112,147]
Pd(ethyl-dtp)$_2$	311	[284]
Pt(ethyl-dtp)$_2$	302	[284]
Co(ethyl-dtp)$_3$	307	[341]
Co(ethyl-dtp)$_2$	313	[341]
Nb(ethyl-dtp)$_4$	356, 274	[127]
V(ethyl-dtp)$_3$	295	[120]

correlation had been suggested by an earlier one [258] between Ni–S distances in planar NiS$_4$ complexes and ligand field strengths taken from the first electronic transition. Additional ligand field strength − metal-sulfur stretching frequency correlations have been discussed [341]. Tables 12 and 13 compare P–S and M–S frequency assignments for a number of R–dtp complexes.

3. Mössbauer Spectra

Very few Mössbauer spectra of R-dtp complexes have been reported. Fe(III)S$_6$ complexes frequently participate in a $^6A_1$ - $^2T_2$ high spin-low-spin equilibrium [373]. Mössbauer studies in conjunction with x-ray emission K_α shifts can be employed [374] to estimate the effective electronic population of iron in complexes. Fe–S bond lengths increase $\sim$ 0.1 Å in passing from the $^2T_2$ to the $^6A_1$ state. Fe(R–dtp)$_3$ complexes are high-spin over a wide range of temperatures [135], e.g., the magnetic moment of Fe(isopropyl-dtp)$_3$ varies from 5.67 to 5.80 B.M. over the temperature range 96.4 – 297.8 °K. Apparently, the only Mössbauer spectra reported [136] are for the compounds Fe(isopropyl-dtp)$_3$ and trans-Fe(isopropyl-dtp)$_2$(pyridine)$_2$.

4. Nuclear Magnetic Resonance Spectra

Very little NMR data has been reported [126,142,143,147,158,193] for the R-dtp complexes. Some data for methyl-dtp complexes is given in Table 14. Generally, the chemical shifts for the complexes are not greatly different than those of related phosphate esters. This has been taken to mean [142] that transmission of d orbital effects via the chelate rings is negligible and that the orbital hybridization employed by phosphorus is essentially constant. The proton magnetic resonance spectrum [143] of Ni(isopropyl-dtp)$_2$ is one characteristic of an isopropoxy group interacting with a phosphorus-31 nucleus. The methyl reso-

Table 14. Proton magnetic resonance data [142]

Compound	Chemical shift, τ	$J_{POCH_3,cps}$
(CH$_3$O)$_3$P=O	6.23	11.0
Ni(Me–dtp)$_2$	6.08	15.6
Pd(Me–dtp)$_2$	6.17	15.6
Zn(Me–dtp)$_2$	6.20	15.8
	a(6.19)	(16)
Cd(Me–dtp)$_2$	6.15	15.7
Pb(Me–dtp)$_2$	6.23	16.0
In(Me–dtp)$_3$	6.03	15.6
Tl(Me–dtp)$^{a)}$	6.33	14

a) Ref. [193].

nance, split into a doublet by the neighboring CH proton, appears at $\delta = 8.62$ with $J_{CH_3} = 6.3$ Hz. The CH absorption appears as a 14-peak multiplet with $\delta = 4.95$ and $J_{CH} = 2.5$ Hz with $J_{POCH} = 6.5$ Hz. In view of conformational studies [375], correlations of Taft σ^* constants with ^{31}P chemical shifts [376] and metal-phosphorus coupling in various complexes [377], it is surprising that the proton and ^{31}P NMR of R-dtp complexes have not received more attention.

By far the most NMR studies have been concerned with the amine adducts of Ni(R-dtp)$_2$ complexes. The green complexes formed with primary (mono- and bidentate) and heterocyclic (mono- und bidentate) amines and the yellow-brown five-co-ordinate adducts with secondary amines exhibit paramagnetic shifts in their NMR spectra. The use of paramagnetic NMR shifts in mapping unpaired electron spin distributions of paramagnetic complexes has been reviewed [378-381]. The upfield paramagnetic shifts of NH protons are diagnostic

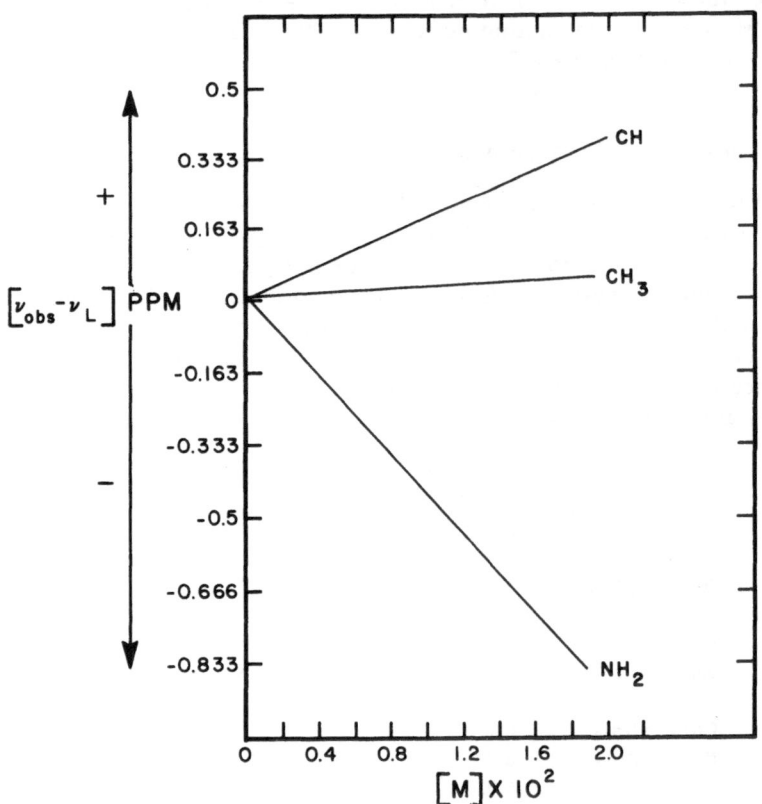

Fig. 5. Shift differences (at 60 MH$_z$) vs. molar concentration of Ni(isobutyl-dtp)$_2$. $T = 310\ ^\circ$K. Solvent: isopropylamine

of NH coordination [322-324,147] and the large paramagnetic shifts enable NiS_4 complexes to have some use as NMR "shift reagents" [322]. In Fig. 5 chemical shift differences are given for neat isopropylamine as a function of the concentration of Ni(isobutyl-dtp)$_2$. It is noted that NH proton absorption is severely broadened upon formation of paramagnetic adducts. This accounts for the NH contact shifts not being reported in the classic paper of Happe and Ward [382] on the amine adducts of cobalt(II) and nickel(II) acetylacetonates. Since NH proton line broadening is a steep function of the concentration of the paramagnetic ion, neat amines must be employed so that the paramagnetic shifts can be observed. A study of the five-co-ordinate adduct of hexamethylphosphoramide (HMPA) with various Ni(R-dtp)$_2$ complexes [324] showed that the ^{31}P NMR spectrum of HMPA was shifted and severely broadened upon adduct formation (Fig. 6). The mechanisms of electron delocalization [143, 147,324,326,383] in the adducts of Ni(R-dtp)$_2$ complexes have been discussed. Fernando and Shetty [384] have shown that it is possible to distinguish Ni[S$_2$P(C$_2$H$_5$)–(OC$_2$H$_5$)]$_2$, Ni[S$_2$P(OC$_2$H$_5$)$_2$]$_2$ and Ni[S$_2$P(C$_2$H$_5$)$_2$]$_2$ using PMR spectroscopy.

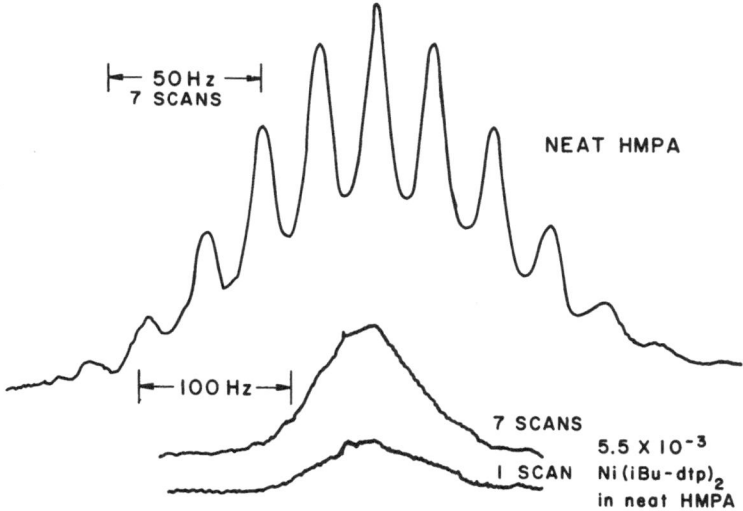

Fig. 6. Phosphorus-31 NMR spectra of neat hexamethylphosphoramide (HMPA) and (lower portion of figure) 5.5×10^{-3} M solution of Ni(isobutyl-dtp)$_2$ in HMPA. The number of scans required to obtain spectra and the spectral widths (Hz) are indicated

5. Electron Spin Resonance Spectra

Ultraviolet irradiation of O, O'-dialkyldithiophosphoric acid glasses at liquid nitrogen temperatures leads to the formation [384] of radical species of the type

$(RO)_2PS_2 \cdot$. The anisotropic ESR spectra were almost identical for all the compounds examined. The signal intensities increased with time during twenty minutes irradiation with little variation in line shape. No other photolytic reactions were detected. The phosphorus hyperfine coupling was almost isotropic in contrast to the g-values. The fractional $3s$ character of the unpaired electrons was estimated to be about 0.7% in the radicals. It was suggested that the extent of spin-polarization and the $3s$ character of phosphorus are nearly the same in dithiophosphate and phosphate (from irradiated PO_4^{3-} salts) radicals. Thiophosphoniumhydrazyl free radicals [385], obtained from the lead(IV) oxide oxidation of $(C_6H_5)_2N-NH-P(=S)R_2$ (where R = C_6H_5, OC_6H_5, OC_2H_5), do not exhibit ^{31}P hyperfine splitting. This indicates that the unpaired electron is located almost completely on the nitrogen from which the hydrogen was removed.

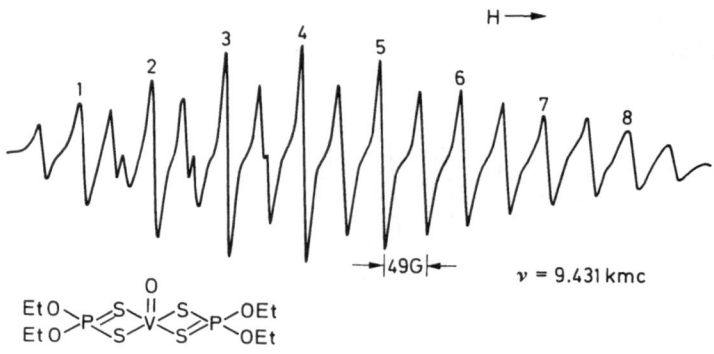

Fig. 7. ESR spectrum of oxobis(0,0'-diethyldithiophosphato)vanadium in benzene. The eight vanadium hyperfine lines are indicated

Oxovanadium(IV) complexes with dithiophosphate ligands have been extensively examined [118,121,161,252,386]. A typical ESR spectrum is shown in Fig. 7. In addition to the eight vanadium ($I = 7/2$) hyperfine lines phosphorus ($I = 1/2$) superhyperfine splitting is also observed. The phosphorus superhyperfine splitting can be considered a bit unusual since the phosphorus is located about 3 Å or more away from the metal ion. ^{31}P and ^{75}As superhyperfine splitting has been observed in the ESR spectra of ill-defined vanadium phosphine [388] and arsine [389] complexes but in those cases, presumably, direct V–P and V–As interactions occur. ESR parameters have been tabulated for a large number of dithiophosphate [121,252] and dithiophosphinate [121,252] complexes. Evaluation [121] of the fractional $3s$ character of unpaired electron in dithiophosphate complexes yielded a value of 1.35%. The vanadyl(IV) complexes possess approximate C_{2v} symmetry. The unpaired "d" electron resides in a $^2A_1$ ground state which is approximately 4.5% of d_{z^2} and 94.5% of $d_{x^2-y^2}$ orbital character as indicated by a charge-consistent extended-Hückel molecular orbital calculation of VO(ethyl-dtp)$_2$ and various crystal field calculations [289]. It is important to note that the vanadium $a_{x^2-y^2}$ orbital is not strongly σ-bonding with respect to sulfur (Fig. 8) but that it does possess the correct

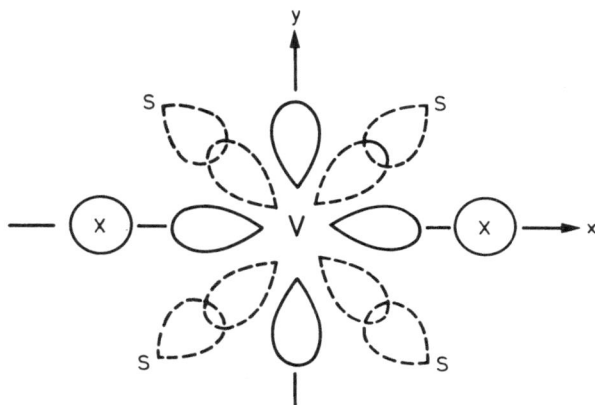

Fig. 8. Direct delocalization mechanism for vanadyl(IV) complexes

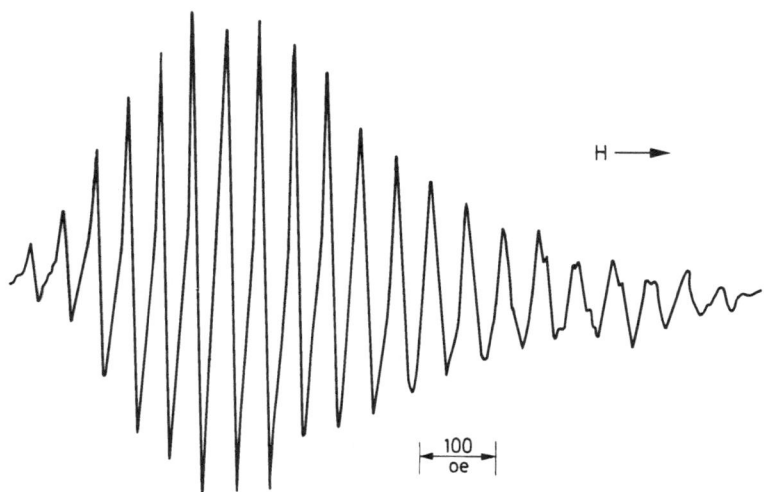

Fig. 9. ESR spectrum of oxobis(dimethyldithioarsinato)vanadium(IV) in benzene

111

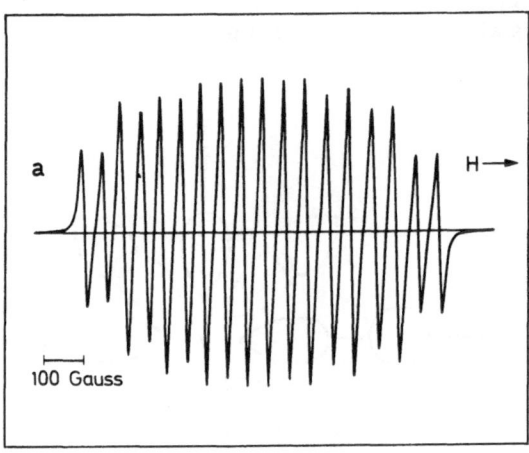

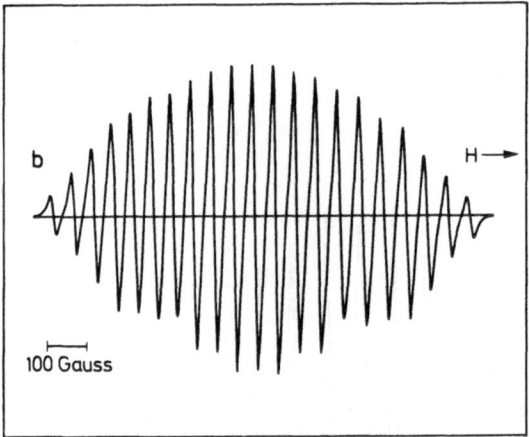

Fig. 10. Simulated ESR spectra of VO(dtcac)$_2$ in benzene.

A) One vanadium nucleus (^{51}V A = 90.63 gauss) and one arsenic nucleus (^{75}As A = 45.31 gauss).

B) One vanadium nucleus (^{51}V A = 90.63 gauss) and two arsenic nuclei (^{75}As A = 45.31 gauss).

A thousand gauss sweep was considered. The spectra were constructed using Lorentzian lines with a peak-to-peak width of 12.3 gauss

112

symmetry to interact directly with phosphorus $3s$ and $3p$ orbitals. This indicates that the *major* source of the large ^{31}P superhyperfine splitting in vanadyl dithiophosphate complexes can be attributed to a direct metal $3d$-phosphorus $3s$-inter-action. Delocalization of the unpaired electron to the phosphorus atom *via* metal-sulfur interaction is also expected to contribute to the observed ^{31}P superhyperfine structure. A similar rationale of the large ^{31}P superhyperfine splitting in vanadyl complexes has been put forward independently by Kozyrev and his co-workers [161]. Recently, it has been possible to show that the preceding delocalization mechanism also applies to chelates with other than phosphorus containing ligands. Oxobis(dithioarsinato)vanadium(IV), $VO[S_2As(CH_3)_2]_2$, $VO(dtcac)_2$, has been found [390] (Fig. 9) to exhibit arsenic-75 superhyperfine splitting. Fig. 10 shows the simulated ESR spectra for vanadyl with one and two $^-S_2 As(CH_3)_2$ bound to the metal. Although not perfect, the simulation is adequate to show that the observed spectrum (Fig. arises from the *bis*-chelate. $VO(ethyl-dtp)_2$ impurities in $V(ethyl-dtp)_3$ are readily detected since large zero-field splittings [391] render the detection of the vanadium(III) species impossible at X-band frequencies.

ESR studies of titanium(III) and zirconium(III) *cy*clopentadienylorganophosphide complexes have been reported [392] but no studies of related dithiophosphate compounds have been described.

The ESR spectra of a few chromium(III) complexes have been described [122,393,394]. The spectra of $Cr(R-dtp)_3$ $(R = CH_3, C_2H_5, n-C_3H_7)$ complexes (powdered samples) consist of a single broad line with a peak-to-peak width of about 300 gauss and a g-value of 1.980 ± 0.010. Assuming effective octahedral symmetry the average g-values for chromium(III) complexes are given by [395]

$$\langle g \rangle = 2.0023(1-4\lambda'/(10Dq))$$

where 10 Dq is the ligand field splitting and λ' is the effective spin-orbital coupling constant for the metal in the complex. $\lambda \cong 40$ cm^{-1}, a value some 56% below the free ion value for $Cr(R-dtp)_3$ complexes. Single crystal ESR sudies [393] of $Cr(ethyl-dtp)_3$ showed the g-value to be close to isotropic $(g = 1.990 \pm 0.001)$ and the zero-field splitting parameters, D and E, to have values of $\pm (138 \pm 1)X 10^{-4}$ cm^{-1} and $\mp (814 \pm 1) X 10^{-4}$ cm^{-1}, respectively. Octrahedral CrS_6 in $CdIn_2S_4$ was found [396] to have $g = 1.995 \pm 0.005$, $g_{\parallel} = 2.000 \pm 0.005$ and $D = - (0.187 \pm 0.002)$ cm^{-1}. A variety of studies of chromium(V), molybdenum(V) and tungsten(V) R-dtp complexes in solution [161,397-399] have been described but the complexes have not been carefully characterized. Comparable studies have also been reported [400] for chromium(V) and molybdenum(V) complexes of dithizon.

An attempt [401] to obtain ESR spectra of frozen toluene solutions of $Mn(ethyl-dtp)_2$ and $Fe(ethyl-dtp)_3$ was unsuccessful. Very large zero-field splittings may have been responsible for the failure of these experiments [401,402]. Preparation of nitrosyl derivatives [403,404] such as Mn(NO)(R-dtp)$_2$ and Fe(NO)(R-dtp)$_2$ may be more productive. The ESR spectra of *tris*(dithioacetylacetonato)iron(III) [405] and manganese(III) dithiocarbamates [406] have been described.

The ESR spectra of copper(II) dithiophosphate complexes have been extensively studied [90,121,161,386,407-418]. Solutions and frozen solutions have been studied exhaustively whereas comparatively little single-crystal data is available [159,387]. A typical solution ESR spectrum is shown in Fig. 11. The high

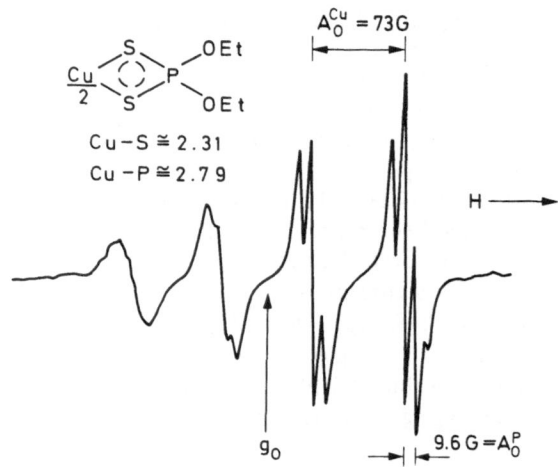

Fig. 11. ESR spectrum of *bis*(0,0′-diethyldithiophosphato)copper(II) in anisole. Estimated structural parameters are also given

field portion of the spectrum definitely shows the effect of the interaction of the unpaired electron on copper with two equivalent phosphorus nuclei. Copper(II)-arsenic and phosphorus interactions have been reported [419] for arsine and phosphine adducts of *bis*(hexafluoroacetylacetonato)copper(II), Cu(Hfac)$_2$. Cu(Hfac)$_2$ interacts with phosphines, *e.g.*, P(C$_6$H$_5$)$_3$, to form 1 : 1 complexes whose ESR spectra [419] exhibit eight lines with ^{31}P$\langle A \rangle \sim 136$ gauss. In these complexes direct (σ) bonding between copper and phosphorus is responsible for the large phosphorus superhyperfine coupling constants. The Cu(R-dtp)$_2$ complexes possess approximate D_{2h} symmetry and the unpaired electron resides in a $^2B_{1g}$ antibonding orbital. In contrast to the vanadyl complexes discussed previously, the copper $3d_{xy}$ orbital is strongly σ-bonding with respect to the sulfur atoms but is π-bonding with respect to the phosphorus $3s$ orbitals (*i.e.*, noninteracting since the overlap integrals vanish by symmetry). The major source of ^{31}P superhyperfine splitting in the ESR spectra is thought [121] to arise from electron delocalization *via* sulfur to the phosphorus atoms. The fractional phosphorus $3s$ character of the ground state of the copper(II) complexes has been estimated [121] to be 0.26%. Weeks and Fackler [159] have found evidence that the g and A tensors are not coaxial in the molecular plane of Cu(ethyl-dtp)$_2$. Other studies [420-424] have shown that complexes with Cu–S and Cu–Se bonds have about the same degree of covalency.

VI. Conclusions

In this review we have assessed the current state of knowledge of the chemistry and properties of ditiophosphate complexes. Although the pioneering work of Jørgensen [132] served to call attention to R-dtp compounds, many additional studies remain to be done. Compounds of many elements in the periodic table have not been characterized and a limited amount of structural data is available. The effect of substituent variation has been shown to affect structures but not to significantly vary electronic properties of complexes with similar structures. It is hoped that this work will assist investigations into the chemistry of dithiophosphate complexes. Although it has not been emphasized, there is an appreciable similarity between dithiophosphinate [9,425,426] and dithiophosphate complexes although significant differences in the properties of the compounds frequently occur.

Before concluding, it is worthwhile to note that photoelectron spectra [427] and X-ray K-absorption spectra [428] have not been reported for R-dtp complexes. The results of these techniques should be particularly useful for characterizing and probing the electronic structure of dithiophosphate complexes.

Acknowledgment. Special thanks are due to Dr. S. L. Lawton, Mobil Research and Development Corporation, Professor B. J. McCormick, West Virginia University, Professor R. G. Cavell, University of Alberta, Professor A. Müller, Lehrstuhl für Anorganische Chemie der Universität Dortmund, Professor W. Kuchen, Institut für Anorganische Chemie der Universität Düsseldorf, Professor T. A. Stephenson, University of Edinburgh, Professor J. P. Fackler, Case-Western Reserve University, Professor A. F. Schreiner, North Carolina State University, and Dr. R. Kirmse, Karl-Marx-Universität, for data prior to publication as well as unpublished data.

115

J. R. Wasson, G. M. Woltermann and H. J. Stoklosa

Addendum

This section has been added in order to bring this review up to date as well as include important references overlooked earlier.

1) Ito, T.: The crystal structure of metal diethyldithiophosphates. II. Lead diethyldithiophosphate. Acta cryst. *B28*, 1034 (1972).
2) Hertel, H., Kuchen, W.: Metallcomplexe der Phosphinsäuren. VI. Über die Elektronenspektren von Thio- und Selenophosphinatokomplexen des Chroms (III). Chem. Ber. *104*, 1735 (1971).
3) Kuchen, W., Mayatepak, H.: Metallcomplexe der Phosphinsäuren. V. Über Dithiosphinatokomplexe von Kupfer, Silber, Gold und Thallium. Chem. Ber. *104*, 3454 (1968).
4) Cole-Hamilton, D. J., Armit, P. W., Stephenson, T. A.: Novel carbonylation products of ruthenium (II) dithioacid complexes. Inorg. Nucl. Chem. Letters *8*, 917 (1972).
5) Alison, J. M. C., Stephenson, T. A.: Metal complexes with sulfur ligands. III. Reaction of platinum(II) N,N-dialkyldithiocarbamates, O-ethyldithiocarbonate(Xanthate) and O, O'-diethyldithiophosphate with tertiary phosphines. J. Chem. Soc. Dalton, in press.
6) Foss, O.: Di-O-alkylmonothiophosphates and di-O-alkylmonoselenophosphates and the corresponding pseudohalogens. Acta Chem. Scand. *1*, 8 (1947).
7) Marshall, T., Fernando, Q.: Thermodynamic constants of the adducts formed with heterocyclic nitrogen bases and nickel(II) dithiophosphates. Anal. Chem. *44*, 1346 (1972).
8) Hazel, J. P., Collin, R. L.: The crystal structure of potassium O,O-dibenzylphosphorodithioate, $KS_2P(O-CH_2-C_6H_5)_2$. Acta cryst *B28*, 2279 (1972).
9) Watanabe, Y., Hagihara, H.: Crystal structures of mercury ethylxanthate and mercury diethyldithiophosphate. Abstracts international union of crystallography, Kyoto, Japan 1972 (Acta Cryst. *B28*, 4 (1972)).
10) Fernando, Q.: Professor on University of Arizona. Personal communication. X-ray crystallographic data reveal that $(Ni(Et-dtp)_2)_2 \cdot DABCO$ (DABCO = 1,4-diazabiclo [2.2.2]octane) contains paramagnetic 5-co-ordinate nickel(II) NiS_4N chromophores.
11) Garif'yanov, N. S., Luchkina, S. A.: EPR study of certain nitrosyl compounds of chromium. Teor. i Eksperim. Khim. Acad. Nauk Ukr. SSR *5*, 571 (1969).
12) Vishnevskaya, G. P., Donskaya, I. S., Karimova, A. F., Kozyrev, B. M.: Electron spin-lattice relaxation in solutions of chromium(III) compounds. Dokl. Akad. Nauk SSSR *202*, 1352 (1972).
13) Yordanov, N. D., Shopov, D.: EPR studies of dithiophosphate and dithiocarbamate complexes. Single-crystal study of bis(O,O'-diisopropyldithiophosphato)copper(II). Chem. Phys. Letters *16*, 60 (1972).
14) Cowsik, R. K., Srinivasan, R.: ESR investigation of copper(II) bis-diethyldithiophosphate in single crystals. Chem. Phys. Letters *16*, 183 (1972).
15) Shopov, D., Yordanov, N. D.: Interaction of copper(II) dithiophosphates and copper(II) dithiocarbamates with organic hydroperoxides-EPR study. Proc. XIVth Intern. Conf. Chem. Toronto *1972*, 236.
16) Cavell, R. G., Sanger, A. R.: Metal complexes with substituted dithiophosphinic acids. VI. Reactions of difluorodithiophosphinic acid with chlorides and oxychlorides of chromium, molybdenum and tungsten. Inorg. Chem. *11*, 2011 (1972).
17) Cavell, R. G., Sanger, A. R.: Metal complexes of substituted dithiophosphinic acids. VII. Reactions of $TiCl_4$, VCl_4, $NbCl_5$ and $TaCl_5$ with difluorodithiophosphinic acid. Inorg. Chem. *11*, 2016 (1972).
18) Michalski, J.: The chemistry and stereochemistry of organic selenium-phosphorus acids and phosphine selenides. Ann. N.Y. Acad. Sci. *192*, 90 (1972).

116

[19] Vincents, H., Schousboe-Jensen, F., Hazell, R. G.: The crystal structure of two modifications of chromium(III) tris(diethyldithiophosphate), $Cr[S_2P(OC_2H_5)_2]_3$. Acta Chem. Scand. 26, 1375 (1972).

[20] Semenov, E. V., Solozhenkin, P. M., Zemlyanskii, N. I., Mel'nik, Y. I.: EPR study of Mo(V) Dtp complexes. Dokl. Akad. Nauk Tadzh SSR 15, 40 (1972); − C. A. 77, 5428m (1972).

[21] Yocn, N., Incervia, M., Zink, J. I.: Paramagnetic 1 : 1 adduct of triphenylphosphine and bis(diethyldithiophosphato)nickel(II). Chem. Commun. 1972, 499.

[22] Toropova, V. F., Cherkasov, R. A., Saveleva, N. I., Slyusar, N. V., Podovik, A. N.: Complexes of phosphorus dithioacids with Ni(II) and Co(II) ions use of the hammett equation with sigma constants for complexing reactions. Zh. Obshch. Khim. 42, 1485 (1972); − C. A. 77, 93581t (1972).

[23] Larin, G. M., Dyatkina, M. E.: ESR of $Cu[S_2P(OC_2H_5)_2]_2$ and $Cu[Se_2P(OC_2H_5)_2]_2$. Izv. Akad. Nauk SSSR Ser. Khim. 1972, 1413; − C. A. 77, 95098q (1972).

[24] Powell, D. B., Scott, J. G.: Infrared spectra of the thiophosphates of tris(ethylenediamine)cobalt(III). Spectrochim. Acta 28A, 1067 (1972).

[25] Golding, R. M., Tennant, W. C.: An ESR study of a copper(II) complex of tetraethylthiuramdisulfide. Mol. Phys. 24, 301 (1972).

[26] Burke, J. M., Fackler, J. P., Jr.: Vibrational spectra of the thiocarbonate complexes of nickel(II) and platinum(II). Inorg. Chem. 11, 2744 (1972).

[27] Schreiner, A. F., Hauser, P. J.: Magnetic circular dichroism spectra and electronic structures of tris(dialkyldithiocarbamato)chromium(III) molecules, $Cr(R_2tc)_3$, and others. Inorg. Chem. 11, 2706 (1972).

[28] Sevdic, D., Meider-Gorican, H.: Solvent extraction of mercury(II) with thiophosphorus compounds. J. Inorg. Nucl. Chem. 34, 2903 (1972).

[29] Zingaro, R. A.: The chemistry of selenium-bearing organometallic derivatives of group VA elements. Ann. N.Y. Acad. Sci. 192, 72 (1972).

[30] Wheatland, D. A., Clapp, C. H., Waldron, R. W.: Complexes of bridged diphosphinothiooyl chelates. Inorg. Chem. 11, 2340 (1972).

[31] Charlton, T. L., Cavell, R. G.: Nuclear magnetic resonance spectra of some oxygen-, sulfur-, and nitrogen-bridged diphosphorus tetrafluoride compounds. Inorg. Chem. 11, 1583 (1972).

[32] Harris, R. K., Woplin, J. R., Murray, M., Schmutzler, R.: Preparation and nuclear magnetic resonance spectra of symmetrical spin systems containing phosphorus: bis(fluorophosphinothioyl) sulfides. J. Chem. Soc. Dalton 1972, 1590.

VII. References

[1] Livingstone, S. E.: Quart. Rev. Chem. Soc. (London) 19, 386 (1965).

[2] Thorn, G. D., Ludwig, R. A.: The dithiocarbamates and related compounds. Amsterdam: Elsevier 1962.

[3] Gray, H. B.: Transition Metal Chem. 1, 240 (1965).

[4] Kimura, T.: Struct. Bonding 5, 1 (1968).

[5] Dräger, M., Gattow, G.: Angew. Chem. Intern. Ed. 7, 868 (1968).

[6] Jørgensen, C. K.: Inorg. Chim. Acta Rev. 2, 65 (1968).

[7] McCleverty, J. A.: Progr. Inorg. Chem. 10, 49 (1968).

[8] Lindoy, L. F.: Coord. Chem. Rev. 4, 41 (1969).

[9] Kuchen, W., Hertel, H.: Angew. Chem. Intern. Ed. 8, 89 (1969).

10) Schrauzer, G. N.: Accts. Chem. Res. *2*, 72 (1969).
11) Coucouvanis, D.: Progr. Inorg. Chem. *11*, 233 (1970).
12) Eisenberg, R.: Progr. Inorg. Chem. *12*, 295 (1970).
13) Tsibris, J. C. M., Woody, R. W.: Coord. Chem. Rev. *5*, 417 (1970).
14) Karayannis, N. M., Mikulski, C. M., Pytlewski, L. L.: Inorg. Chim. Acta Rev. *5*, 69 (1971).
15) Livingstone, S. E.: Coord. Chem. Rev. *7*, 59 (1971).
16) Cox, M., Darken, J.: Coord. Chem. Rev. *7*, 29 (1971).
17) Holm, R. H., O'Connor, M. J.: Progr. Inorg. Chem. *14*, 241 (1971).
18) Schrauzer, G. N.: Transition Metal Chem. *4*, 299 (1968).
19) Spence, J. T.: Coord. Chem. Rev. *4*, 475 (1969).
20) Furlani, C., Luciani, M. L.: Inorg. Chem. *7*, 1586 (1968).
21) Melson, G. A., Crawford, N. P., Geddes, B. J.: Inorg. Chem. *9*, 1123 (1970).
22) Bonamico, M., Dessy, G., Fares, V.: Chem. Commun. *1969*, 324.
23) Bonamico, M., Dessy, G., Fares, V., Porta, P., Scaramuzza, L.: Chem. Commun. *1971*, 365.
24) Fackler, J. P.: J. Am. Chem. Soc. *94*, 1009 (1972).
25) Furlani, C., Piovesana, O., Tomlinson, A. A. G.: J. C. S. Dalton *1972*, 212.
26) Stiddard, M. H. B., Townsend, R. E.: J. Chem. Soc. A *1970*, 2720.
27) Fenn, R. H., Segrott, G. R.: J. Chem. Soc. A *1970*, 2781, 3197.
28) Cervone, E., Camassei, F. D., Luciani, M. L., Furlani, C.: J. Inorg. Nucl. Chem. *31*, 1101 (1969).
29) Battistoni, C., Mattogno, G., Monaci, A., Tarli, F.: J. Inorg. Nucl. Chem. *33*, 3815 (1971).
30) Battistoni, C., Mattogno, G., Monaci, A., Tarli, F.: Inorg. Nucl. Chem. Letters *7*, 1081 (1971).
31) Ali, M. A., Livingstone, S. E., Phillips, D. J.: Inorg. Chim. Acta *6*, 39 (1972) and references therein.
32) Coucouvanis, D., Lippard, S. J.: J. Am. Chem. Soc. *90*, 3281 (1968).
33) Fries, D. C., Fackler, J. P.: Chem. Commun. *1971*, 276.
34) Flamini, A., Furlani, C., Piovesana, O.: J. Inorg. Nucl. Chem. *33*, 1841 (1971).
35) Fackler, J. P., Coucouvanis, D., Fetchin, J. A., Siedel, W. C.: J. Am. Chem. Soc. *90*, 2784 (1968).
36) Giuliani, A. M.: Inorg. Nucl. Chem. Letters *7*, 1001 (1971).
37) Hart, D. M., Rolfs, P. S., Kessinger, J. M.: J. Inorg. Nucl. Chem. *32*, 469 (1970).
38) Fabretti, A. C., Pellacani, G. C., Peyronel, G.: J. Inorg. Nucl. Chem. *33*, 4247 (1971).
39) Pellacani, G. C., Peyronel, G.: Inorg. Nucl. Chem. Letters *8*, 299 (1972).
40) Pellacani, G. C., Feltri, T.: Inorg. Nucl. Chem. Letters *8*, 325 (1972).
41) Musker, W. K., Hill, N. L.: Inorg. Chem. *11*, 710 (1972).
42) Baker, D. J., Goodall, D. C., Moss, D. S.: Chem. Commun. *1969*, 325.
43) Steger, H. F.: J. Inorg. Nucl. Chem. *33*, 3399 (1971).
44) Podlaha, J., Podlahova, J.: Inorg. Chim. Acta *5*, 413, 420 (1971) and references therein.
45) Herskovitz, T., Forbes, C. E., Holm, R. H.: Inorg. Chem. *11*, 1318 (1972) and references therein.
46) Davison, A., Switkes, E. S.: Inorg. Chem. *10*, 837 (1971).
47) Churchill, M. R., Cooke, J., Fennessey, J. P., Wormald, J.: Inorg. Chem. *10*, 1031 (1971).
48) Churchill, M. R., Wormald, J.: Inorg. Chem. *10*, 1778 (1971).
49) Churchill, M. R., Cooke, J., Wormald, J., Davison, A., Switkes, E.: J. Am. Chem. Soc. *91*, 6518 (1969).

[50] Girling, R. L., Amma, E. L.: Chem. Commun. *1968*, 1487.
[51] Luth, H., Hall, E. A., Spofford, W. A., Amma, E. L.: Chem. Commun. *1969*, 520 and references therein.
[52] Dwyer, F. P., Sargeson, A. M.: J. Am. Chem. Soc. *81*, 2335 (1959).
[53] Carlin, R. L., Canziani, F.: J. Chem. Phys. *40*, 371 (1964).
[54] Coucouvanis, D., Coffman, R. E., Piltingsrud, D.: J. Am. Chem. Soc. *92*, 5004 (1970).
[55] Sweeney, W. V., Coffman, R. E.: J. Phys. Chem. *76*, 49 (1972).
[56] Ojima, I., Iwamoto, T., Onishi, T., Inamoto, N., Tamaru, K.: Chem. Commun. *1969*, 1501.
[57] Ojima, I., Iwamoto, T., Onishi, T., Inamoto, N., Tamaru, K.: Bull. Chem. Soc. Japan *44*, 2150 (1971).
[58] Reiff, O. M., Andress, H. J.: U. S. 2,438,876, March 30, 1948; Chem. Abstr. *42*, 5657a (1948).
[59] Funk, C. E.: U. S. 2,466,408, April 5, 1949; Chem. Abstr. *43*, 4845d (1949).
[60] Funk, C. E.: Brit. 636,552, May 3, 1950; Chem. Abstr. *44*, 9668d (1950).
[61] Funk, C. E.: U. S. 2,535,024, December 26, 1950; Chem. Abstr. *45*, 2194c (1951).
[62] McDermott, J. P.: U. S. 2,529,303, November 7, 1950; Chem. Abstr. *45*, 1763c (1951).
[63] Vold, M. J.: U. S. 2,528,257, October 31, 1950; Chem. Abstr. *45*, 1339b (1951).
[64] McNab, J. G., Hakala, N. V., McDermott, J. P.: U. S. 2,552,570, May 15, 1951; Chem. Abstr. *45*, 7785e (1951).
[65] Rudel, H. W., Kirshenbaum, A. D.: U. S. 2,595,170, April 29, 1952; Chem. Abstr. *46*, 7318g (1952).
[66] Evans, E. A., Elliot, J. S.: U. S. 2,579,037, December 18, 1951; Chem. Abstr. *46*, 2795b (1952).
[67] Neely, R. J., Gilmartin, R. P.: U. S. 2,678,262, May 11, 1954; Chem. Abstr. *48*, 9676d (1954).
[68] Neely, R. J., Gilmartin, R. P.: Brit. 684,155, December 10, 1952; Chem. Abstr. *48*, 1425d (1954).
[69] Hersh, J. M.: U. S. 2,688,013, August 31, 1954; Chem. Abstr. *49*, 606c (1955).
[70] Mulvany, P. K.: U. S. 2,689,220, September 14, 1954; Chem. Abstr. *49*, 605e (1955).
[71] Heisig, T. C., Murphey, R. L.: U. S. 2,710,842, June 14, 1955; Chem. Abstr. *49*, 13,641f (1955).
[72] Bartleson, J. D.: U. S. 2,794,713, June 4, 1957; Chem. Abstr. *51*, 13,380b (1957).
[73] Wystrach, V. P., Hook, E. O., Christopher, G. L.: U. S. 2,794,780, June 4, 1957; Chem. Abstr. *51*, 13379e (1957).
[74] Brugmann, W. H.: U. S. 2,786,029, March 19, 1957; Chem. Abstr. *51*, 13,379g (1957).
[75] Gilbert, L. F.: U. S. 2,794,717, June 4, 1957; Chem. Abstr. *51*, 13,379f (1957).
[76] Scanley, C. S.: U. S. 2,799,653, July 16, 1957; Chem. Abstr. *51*, 15, 112i (1957).
[77] Smith, H. M., Hoerrner, W., Davidson, J. R.: U. S. 2,824,836, February 25, 1958; Chem. Abstr. *52*, 9583d (1958).
[78] McDermott, J. P.: U. S. 2,766,207, October 9, 1956; Chem. Abstr. *52*, 9584d (1958).
[79] Alford, H. E., Liao, C.-W.: U. S. 2,841,551, July 1, 1958; Chem. Abstr. *52*, 19,112g (1958).
[80] Reeves, R. F., Cestoni, D. J.: U.S. 2,837, 589, June 3, 1958; Chem. Abstr. *52*, 19,112d (1958).
[81] Goldsmith, F. C.: U. S. 2,838,555, June 10, 1958; Chem. Abstr. *52*, 15,896g (1958).
[82] Butler, T. A.: U. S. 2,861,907, November 25, 1958; Chem. Abstr. *53*, 11,183i (1959).
[83] Higgins, W. A., Wurstner, R. G.: U. S. 2,987,410, Appl. March 13, 1958; Chem. Abstr. *56*, 6264h (1962).

119

84) Lynch, C. S., Lifson, W. E., Finn, R. F.: U. S. 3,014,940, December 26, 1961; Chem. Abstr. 56, 13,164d (1962).
85) Lynch, C. S., Lifson, W. E., Finn, R. F.: Brit. 876,505, Appl. May 20, 1959; Chem. Abstr. 56, 14,082b (1962).
86) Kreutzer, I., Ludwig, G.: Ger. Offen. 2,020,454, November 11, 1971; Chem. Abstr. 76, 33,791f (1972).
87) Kreutzer, I., Ludwig, G.: Ger. Offen. 2,020,455, November 11., 1971; Chem. Abstr. 76, 33,792g (1972).
88) Vipper, A. B., Papok, K. K., Sanin, P. I., Sher, V. V.: Khim. i Tekhnol. Topliv. i Masel 3, 45 (1958); – Chem. Abstr. 52, 12381f (1958).
89) Sanin, P. I., Sher, V. V., Nikitskaya: Khim i Tekhnol. Topliv. i Masel 3, 24 (1958); – Chem. Abstr. 52, 21,020e (1958).
90) Shimonaev, G. S., Zakharov, G. V.: Khim. i Tekhnol. Topliv. i Masel 11, 59 (1966); – Chem. Abstr. 65, 19, 897a (1966).
91) Larson, R.: Erdoel Kohle 11, 791 (1958); – Chem. Abstr. 53, 6590g (1959).
92) Bennett, P. A.: S. A. E. J. 1958, 10713; – Chem. Abstr. 54, 21, 731 c (1960).
93) Scanley, C. S., Larson, R.: S. A. E. J. 1958, 107c; – Chem. Abstr. 54, 21, 731f (1960).
94) Shapov, D., Ivanov, S.: Izv. Inst. Obshcha Neorg. Khim., Bulgar. Akad. Nauk. 8, 239 (1961); – Chem. Abstr. 57, 1167g (1962).
95) Monastyrskii, V. N., Fufaev, A. A., Perel'miter, M. S.: Prisadki Maslam i Toplivam, Tr. Nauchn.-Tekhn. Soveshch. 1960, 128: – Chem. Abstr. 57, 1167i (1962).
96) Bencze, P.: Magy. Asvanyolaj Foldgaz Kiserl. Int. Kozlemen. 1, 178 (1948–59) (Publ. 1960); – Chem. Abstr. 57, 2493h (1962).
97) Hoock, W. S., Kleinholz, M. P.: S. A. E. Preprint, 1961, 300B; – Chem. Abstr. 60, 3924a (1964).
98) In'kova, N. M., Piyunkina: Nefteprererabotka i Neftekhim., Nauchn.-Tekhn. Sb. 1968, 11; – Chem. Abstr. 63, 17, 754b (1965).
99) Drummond, A.: Tech. Petrole 20, 9 (1965); – Chem. Abstr. 67, 55,891e (1967).
100) Famer, H. H., Malone, B. W., Tompkins: Lubrication Eng. 23, 57 (1967); – Chem. Abstr. 66, 97,175j (1967).
101) Goodwin, M. C., Begeman, C. R.: S. A. E. Preprint 1962, 458IG; – Chem. Abstr. 59, 11,159d (1963).
102) Geldern, L.: Erdoel Kohle 18, 545 (1965); – Chem. Abstr. 63, 9713h (1965).
103) Murata, N., Okutsu, M.: Junkatsu 12, 286 (1967); – Chem. Abstr. 69, 113, 150s (1968).
104) Analytical methods for pesticides. Plant growth regulators. and food additives. New York: Academic Press, a series of volumes.
105) Hartley, G. S., West, T. F.: Chemicals for pest control. New York: Pergamon Press 1969.
106) Graham: Analytical methods for pesticides. Plant growth regulators, and food additives. Vol. II, p. 223. New York: Academic Press 1964.
107) Hill, A. C.: J. Sci. Food Agr. 20, 4 (1969); – Chem. Abstr. 70, 56,624u (1969).
108) Hill, A. C.: Ger. 927,092, April 28, 1955; – Chem. Abstr. 50, 2653 (1956).
109) Fusco, R., Losco, G., Perini, M.: U. S. 2,901,481, August 25, 1959; – Chem. Abstr. 54, 1297f (1960).
110) Gaher, S., Drabek, J., Truchlik, S., Sirota, T., Batora, V.: Ger. 1,290,008, 27 (1969); – Chem. Abstr. 71, 2965h (1969).
111) Kubo, H.: Agr. Biol. Chem. 29, 43 (1965); – Chem. Abstr. 63, 7032b (1965).
112) Livingstone, S. E., Mihkelson: Inorg. Chem. 9, 2545 (1970).
113) Stary, J.: The solvent extraction of metal chelates. New York: The Macmillan Co.1964.

114) De, A. K., Khopkar, S. M., Chalmers, R. A.: Solvent extraction of metals. New York: Van Nostrand-Reinhold Co. 1970.
115) Zolotov, Yu. A.: Extraction of chelate compounds. Michigan: Ann Arbor-Humphrey Science Publishers, Ann Arbor 1970.
116) Kabovskii, I. A.: Proc. Intern. Congr. Surface Activity, 2nd, London 1957, 225–37; – Chem. Abstr. 54, 2105i (1960).
117) Kabovskii, I. A.: Czech. 95,487, June 15, 1960; – Chem. Abstr. 55, 9242a (1961).
118) Furlani, C., Tomlinson, A.'A. G., Porta, P., Sgamelotti, A.: J. Chem. Soc. A 1970, 2929; – Chem. Commun. 1969, 1969.
119) Lebedda, J. D., Palmer, R. A.: Inorg. Chem. 10, 2704 (1971).
120) Cavell, R. G., Day, E. D., Byers, W., Watkins, P. M.: Inorg. Chem. 10, 2716 (1971).
121) Wasson, J. R.: Inorg. Chem. 10, 1531 (1971).
122) Wasson, J. R., Wasson, S. J., Woltermann, G. M.: Inorg. Chem. 9, 1576 (1970).
123) Cavell, R. G., Byers, W., Day, E. D.: Inorg. Chem. 10, 2710 (1971).
124) Goldberg, D. E., Fernelius, W. C., Shamma, M.: Inorg. Syn. 6, 142 (1960).
125) Cancellieri, P., Cervone, E., Furlani, C., Sartori, G.: Z. Phys. Chem. N.F. 62, 35 (1968).
126) Pantaleo, D. C., Johnson, R. C.: Inorg. Chem. 10, 1298 (1971).
127) McGinnis, R. N., Hamilton, J. B.: Inorg. Nucl. Chem. Letters 8, 245 (1972).
128) Spengler, G., Weber, A.: Chem. Ber. 92, 2163 (1959).
129) Lindoy, L. F., Livingstone, S. E., Lockyer, T.: Australian J. Chem. 18, 1549 (1965).
130) Jowitt, R. N., Mitchell, P. C. H.: J. Chem. Soc. A 1969, 2632.
131) Jowitt, R. N., Mirchell, P. C. H.: J. Chem. Soc. A 1970, 1702.
132) Jørgensen, C. K.: J. Inorg. Nucl. Chem. 24, 1571 (1962).
133) Hartman, F. A., Wojcicki, A.: Inorg. Nucl. Chem. Letters 2, 303 (1966).
134) Malatesta, L., Pizzotti, R.: Chim. Ind. (Milan) 27, 6 (1945); – Chem. Abstr. 40, 7039–6 (1946).
135) Ewald, A. H., Martin, R. L., Sinn, E., White, A. H.: Inorg. Chem. 8, 1837 (1969).
136) Korecz, L., Burger, K., Jørgensen, C. K.: Helv. Chim. Acta 51, 211 (1968).
137) Jørgensen, C. K.: Acta Chem. Scand. 16, 1048 (1962).
138) Jørgensen, C. K.: Acta Chem. Scand. 16, 2017 (1962).
139) Hillis, J. E., DeArmond, M. K.: Chem. Phys. Let. 10, 325 (1971).
140) Mitchell, R. W., Ruddick, J. D., Wilkinson, G.: J. Chem. Soc. A 1971, 3224.
141) Araneo, A., Bonati, F., Minghetti, G.: Inorg. Chim. Acta 4, 61 (1970).
142) Woltermann, G. M., Wasson, J. R.: Inorg. Nucl. Chem. Letters 6, 475 (1970).
143) Francis, H. E., Tincher, G. L., Wagner, W. F., Wasson, J. R., Woltermann, G. M.: Inorg. Chem. 10, 2620 (1971).
144) Nanjo, M., Yamasaki, T.: J. Inorg. Nucl. Chem. 32, 2411 (1970).
145) Wasson, J. R., Angus, J. R., Woltermann, G. M.: Unpublished results.
146) Angus, J. R.: M. S. thesis. Lexington: University of Kentucky, August 1971.
147) Angus, J. R., Wasson, J. R.: J. Coord. Chem. 1, 309 (1971).
148) Angus, J. R., Woltermann, G. M., Wasson, J. R.: J. Inorg. Nucl. Chem. 33, 3967 (1971).
149) Lebedda, J. D., Palmer, R. A.: Inorg. Chem. 11, 484 (1972).
150) Francis, H. E., Tincher, G. L., Wagner, W. F., Wasson, J. R.: Unpublished results.
151) Sgamellotti, A., Furlani, C., Magrini, F.: J. Inorg. Nucl. Chem. 30, 2655 (1968).
152) Ripan, R., Mirel, C., Lupu, D.: Rev. Roumaine Chim. 13, 303 (1968).
153) Carlin, R. L., Dubnoff, J. S., Huntress, W. T.: Proc. Chem. Soc. 1964, 228.
154) Ciullo, G., Furlani, C., Sestili, L., Sgamellotti, A.: Inorg. Chim. Acta 5, 489 (1971).
155) Jørgensen, C. K.: Acta Chem. Scand. 17, 533 (1963).
156) Dakternieks, D. R., Graddon, D. P.: Australian J. Chem. 24, 2509 (1971).
157) Tomlinson, A. A. G., Furlani, C.: Inorg. Chim. Acta 3, 487 (1969).

121

J. R. Wasson, G. M. Woltermann and H. J. Stoklosa

158) Carlin, R. L., Losee, D. B.: Inorg. Chem. 9, 2087 (1970).
159) Weeks, M., Fackler, J. P.: Personal communication.
160) Weeks, M. J., Fackler, J. P.: Inorg. Chem. 7, 2548 (1968).
161) Obchinnikov, I. V., Gainulin, I. F., Garig'yanov, N. S., Kozyrev, B. M.: Dokl. Akad. Naukk SSSR 191, 395 (1970).
162) Knox, J. R., Prout, C. K.: Acta Cryst. B25, 2281 (1969).
163) McConnell, J. F., Kastalsky: Acta Cryst. 22, 853 (1967).
164) Fernando, Q., Green, C. D.: J. Inorg. Nucl. Chem. 29, 647 (1967).
165) Gilinskaya, E. A., Porai-Koshits, M. A.: Kristallografiya 4, 241 (1959).
166) Craig, D. C., Pallister, E. T., Stephenson, N. C.: Acta Cryst. B27, 1163 (1971).
167) Ooi, S., Fernando, Q.: Inorg. Chem. 6, 1558 (1967).
168) Shetty, P. S., Ballard, R. E., Fernando, Q.: Chem. Commun. 1969, 717; see also Ref. 170)
169) Ooi, S., Carter, D., Fernando, Q.: Progress in coordination chemistry, Proc. 11th Internat. Conf. Coord. Chem., Haifa and Jerusalem 1968, (ed. M. Cais). New York: Elsevier Publishing Co. 1968, D43; — Shiro, M., Fernando, Q.: Chem. Commun. 1971, 350; see also Ref. 170)
170) Shetty, P. S., Fernando, Q.: J. Am. Chem. Soc. 92, 3964 (1970).
171) Schrauzer, G. N., Mayweg, V. P., Heinrich, W.: Inorg. Chem. 4, 1615 (1965).
172) Vincent, W. R., Wasson, J. R.: Unpublished results.
172a) Zemlyanskii, N. I., Kalashnikov, V. P., Yarimovich, V. K.: Zh. Obshch. Khim. 39, 1591 (1969).
173) Dickert, J. J., Rowe, C. N.: J. Org. Chem. 32, 647 (1967); see also Ref. 182)
174) Lawton, S. L., Rohrbaugh, W. J., Kokotailo, G. T.: Inorg. Chem. 11, 612 (1972).
175) Lawton, S. L., Rohrbaugh, W. J., Kokotailo, G. T.: Inorg. Chem. 11 (1972) in press.
176) Dickert, J. J., Rowe, C. N.: U. S. Patent No. 3,554,908 (1971); — Chem. Abstr. 74, 55,937w (1971).
177) Ito, T., Igarashi, Hagihara, H.: Acta Cryst. B25, 2303 (1969).
178) Lawton, S. L., Kokotailo, G. T.: Inorg. Chem. 8, 2410 (1969).
179) Lawton, S. L.: Inorg. Chem. 10, 328 (1971).
180) Lawton, S. L., Kokotailo, G. T.: Nature 221, 550 (1969).
181) Lawton, S. L., Kokotailo, G. T.: Inorg. Chem. 11, 363 (1972).
182) Burn, A. J., Smith, G. W.: Chem. Commun. 1965, 394.
183) Bacon, W. E., Bork, J. F.: J. Org. Chem. 27, 1484 (1962); see also Ref. 182)
184) Colclough, T., Cunneen, J. I.: J. Chem. Soc. 1964, 4790.
185) Dakternieks, D. R., Graddon, D. P.: Australian J. Chem. 23, 1989, 2521 (1970).
186) Francis, S. A., Ellison, A. H.: J. Chem. Eng. Data 6, 83 (1961).
187) Hanneman, W. W., Porter, R. S.: J. Org. Chem. 29, 2996 (1964).
188) Dunn, J. R., Scanlan, J.: J. Polymer Sci. 35, 267 (1959).
189) Rockett, J.: Appl. Spectry. 16, 39 (1962).
190) Ashford, J. S., Bretherick, Gould, P.: J. Appl. Chem. 15, 170 (1965).
191) Wystrach, V. P., Hook, E. O., Christopher, G. L. M.: J. Org. Chem. 21, 705 (1956).
192) Coggan, P. L., Lebedda, J. D., McPhail, A. T., Palmer, R. A.: Chem. Commun. 1970, 78.
193) Bonati, F., Minghetti, G.: Inorg. Chim. Acta 3, 161 (1969).
194) Husebye, S.: Acta Chem. Scand. 20, 2007 (1966).
195) Husebye, S., Helland-Madsen, G.: Acta Chem. Scand. 23, 1398 (1969).
196) Wasson, J. R.: Methodicum chimicum Houben-Weyl. Stuttgart: G. Thieme Verlag, in press.
197) Husebye, S.: Acta Chem. Scand. 19, 1045 (1965).
198) Husebye, S.: Acta Chem. Scand. 20, 24 (1966).

199) Pishchimuka, P.: J. Russ. Phys. Chem. *44*, 1406 (1912).
200) Mastin, T. W., Norman, G. R., Weilmuenster: J. Am. Chem. Soc. *67*, 1662 (1945).
201) Fletcher, J. H., Hamilton, J. C., Hechenbleikner, I., Hoegberg, E. I., Sertl, B. J., Cassaday, J. T.: J. Am. Chem. Soc. *72*, 2461 (1961); see also: (ed. Müller, E.) Methoden der Organischen Chemie (Houben-Weyl), Vol. XII, Part 2, pp. 683–690. Stuttgart: Georg Thieme 1964.
202) Bode, H., Arnswald, W.: Z. Anal. Chem. *185*, 99, 179 (1962).
203) Lambert, R. L., Manuel, T. A.: Inorg. Chem. *5*, 1287 (1966).
204) Imaev, M. G.: Zh. Obshch. Khim. *35*, 1864 (1965); – Chem. Abstr. *64*, 1991h (1966).
205) Bengze, P., Baboczky-Kampos, K.: Acta Chim. Hung. *31*, 53 (1962).
206) Orudzheva, I. M., Novruzov, S. M.: Azerb. Neft. Khoz. *48*, 41 (1969); – Chem. Abstr. *71*, 126,851m (1969).
207) Tishkova, V. N., Isagulyants, V. I., Chang, H.-C., Utsmieva, N. M.: Prisadki k Maslam i Toplivam, Tr. Nauchn.-Tekhn. Soveshch. *1960*, 34; – Chem. Abstr. *57*, 1168g (1962).
208) See 207) and Brit. 815,965, July 1, 1959; Chem. Abstr. *54*, 1297g (1960).
209) Durr, A. M.: U. S. 3,210,275, October 5, 1965; Chem. Abstr. *63*, 17,970a (1965).
210) Mulvany, P. K.: U. S. 2,680,123, June 1 (1954); Chem. Abstr. *50*, 2653g (1956).
211) Kalmutchi, V., Musat, T., Panait, I.: Petrol Gaze *19* (Suppl.), 21 (1967); – Chem. Abstr. *70*, 19,705a (1969).
212) Verleg, G. M.: U. S. 2,838,557, June 10, 1958; Chem. Abstr. *52*, 16,197d (1958).
213) Makens, R. F., Vaughan, H. H., Chelberg, R. R.: Anal. Chem. *27*, 1062 (1955).
214) Tseng, C. K., Chan, J. H.-H.: Tetrahedron Letters *1971*, 699.
215) Zemlyanskii, N. I., Drach, B. S.: Zh. Obshch. Khim. *32*, 1962 (1962).
216) Zemlyanskii, N. I., Chernaya, N. M.: Ukr. Khim. Zh. *33*, 182 (1967); – Chem. Abstr. *67*, 7571d (1967).
217) Arbuzov, A. E., Shapshinskaya, O. M.: Tr. Kazansk. Khim. Tekhnol. Inst. i. S. M. Kirova *1951*, 3; – Chem. Abstr. *51*, 5688b (1957).
218) Malatesta, L.: Gazz. Chim. Ital. *81*, 596 (1951).
219) Müller, A., Christophliemk, P., Krishna Rao, V. V.: Chem. Ber. *104*, 1905 (1971).
220) Hu, P.-F., Chen, W.-Y.: Hua Hsueh Hsueh Pao *22*, 215 (1956); – Chem. Abstr. *52*, 7186c (1958).
221) Hu, P.-F., Cheng, W.: Sci. Sinica *6*, 661 (1957); – Chem. Abstr. *52*, 7186h (1958).
222) Herriott, A. W.: J. Am. Chem. Soc. *93*, 3304 (1971).
223) Larionov, S. V., Il'ina, L. A.: Zh. Obshch. Khim. *39*, 1587 (1969); – Chem. Abstr. *71*, 108,556r (1969).
224) Handley, T. H.: Anal. Chem. *35*, 991 (1963) and references therein.
225) Muratova, A. A., Yarkova, E. G., Kuramshin, I. Ya., Pudovik: Zh. Obshch. Khim. *41*, 1668 (1971); – Chem. Abstr. *76*, 20,793h (1972).
226) Rowe, C. N., Dickert, J. J.: ASLE Trans. *10*, 85 (1967).
227) Lawton, S. L.: Inorg. Chem. *9*, 2269 (1970).
228) Galsbøl, F., Schäffer, C.: Inorg. Syn. *10*, 42 (1967).
229) Kudchadker, M. V., Zingaro, R. A., Irgolic, K. J.: Can. J. Chem. *46*, 1415 (1968).
230) Melton, R. G., Zingaro, R. A.: Can. J. Chem. *46*, 1425 (1968).
231) Krishnan, V., Zingaro, R. A.: Inorg. Chem. *8*, 2337 (1969).
232) Krishnan, V., Zingaro, R. A.: J. Coord. Chem. *1*, 1 (1971).
233) Jørgensen, C. K.: Mol. Phys. *5*, 485 (1962).
234) Tomlinson, A. A. G.: J. Chem. Soc. A *1971*, 1409.
235) Corbridge, D. E. C.: Topics Phosphorus Chem. *3*, 57 (1966).
236) Coppens, P., MacGillavry, C. H., Hovenkamp, S. G., Douwes, H.: Acta Cryst. *15*, 765 (1962).

J. R. Wasson, G. M. Woltermann and H. J. Stoklosa

237) Mighell, A. D., Smith, J. P., Brown, W. E.: Acta Cryst. *B25*, 776 (1969).
238) van Houten, S., Wiebenga, E. H.: Acta Cryst. *10*, 156 (1957).
239) Hunt, G. W., Cordes, A. W.: Inorg. Chem. *10*, 1935 (1971).
240) Lee, J. D., Goodacre, G. W.: Acta Cryst. *B27*, 1055 (1971).
241) Saenger, W., Eckstein, F.: J. Am. Chem. Soc. *92*, 4712 (1970).
242) Mootz, D., Goldmann, J.: Acta Cryst. *B25*, 1256 (1969).
243) Bonamico, M., Mazzone, G., Vaciago, A., Zambonelli, L.: Acta Cryst. *19*, 898 (1965).
244) Fraser, K. A., Harding, M. M.: Acta Cryst. *22*, 75 (1967).
245) Domenicano, A., Torelli, L., Vaciago, A., Zambonelli, L.: J. Chem. Soc. A *1968*, 1351.
246) Calligaris, M., Ciana, A., Meriani, S., Nardin, G., Randaccio, L., Ripamonti, A.: J. Chem. Soc. A *1970*, 3386.
247) Calligaris, M., Nardin, G., Ripamonti, A.: J. Chem. Soc. *1970*, 714.
248) Bonamico, M., Dessy, G.: J. Chem. Soc. A *1971*, 264.
249) Lepicard, G., DeSaint-Giniez-Liebig, D., Laurent, A., Rerat, C.: Acta Cryst. *B25*, 617 (1969).
250) Husebye, S.: Acta Chem. Scand. *20*, 51 (1966).
251) Bennett, M. J., Sumner, R.: University of Alberta, Edmonton, Alberta, Canada; work cited in Ref. 252)
252) Cavell, R. G., Day, E. D., Byers, W., Watkins, P. M.: Inorg. Chem. *11*, 1591 (1972).
253) Eisenberg, R., Gray, H. B.: Inorg. Chem. *6*, 1844 (1967).
254) Piovesana, O., Cappuccilli: Inorg. Chem. *11*, 1543 (1972).
255) Gardner, R. A., Vlasse, M., Wold, A.: Acta Cryst. *B25*, 781 (1969).
256) Beurskens, P. T., Cras, J. A., Noordik, J. H., Spruijt, A. M. V.: J. Cryst. Mol. Struct. *1*, 93 (1971).
257) Porta, P., Sgamellotti, A., Vinciguerra, N.: Inorg. Chem. *10*, 541 (1971).
258) Porta, P., Sgamellotti, A., Vinciguerra, N.: Inorg. Chem. *7*, 2525 (1968).
259) Kastalsky, V., McConnell, J. F.: Acta Cryst. *B25*, 909 (1969).
260) Carter, D. E.: Structure and properties of certain planar metal chelates and their adducts, University of Arizona, Ph. D. thesis, 1969. Available from University Microfilms, Inc., Ann Arbor, Michigan, Order No. 69–7618.
261) Johns, P. E., Ansell, G. B., Katz, L.: Chem. Commun. *1968*, 78; – Acta Cryst. *B25*, 1939 (1969).
262) Shetty, P. S., Fernando, Q.: Acta Cryst. *B25*, 1294 (1969).
263) Khare, G. P., Schultz, A. J., Eisenberg, R.: J. Am. Chem. Soc. *93*, 3597 (1971).
264) Cavalca, L., Nardelli, M., Fava, G.: Acta Cryst. *15*, 1139 (1962).
265) Gaspari, G. F., Nardelli, M., Villa, A.: Acta Cryst. *23*, 384 (1967).
266) Bonamico, M., Dessy, G., Mariani, C., Vaciago, A., Zambonelli, L.: Acta Cryst. *19*, 619 (1965).
267) Peyronel, G., Pignedoli, A.: Acta Cryst. *23*, 298 (1967).
268) Grim, S. O., Plastas, H. J., Huheey, C. L., Huheey, J. E.: Phosphorus *1*, 61 (1971); – see also: Cruickshank, D. W. J.: J. Chem. Soc. *1961*, 5486; – Tsvetkov, E. N., Bochvar, D. A., Kabachnik, M. I.: Teor. i Eksperim. Khim. *3*, 3 (1967).
269) Keeton, M., Santry, D. P.: Chem. Phys. Letters *7*, 105 (1970).
270) Urch, D. S.: J. Chem. Soc. A *1969*, 3026.
271) Gianturco, F. A.: J. Chem. Soc. *1969*, 1293.
272) Maclagan, R. G. A.: J. Chem. Soc. A *1971*, 222.
273) Maclagan, R. G. A.: J. Chem. Soc. A *1970*, 2992.
274) Marsmann, H., Van Wazer, J. R., Robert, J. B.: J. Chem. Soc. A *1970*, 1566.
275) Mitchell, K. A. R.: Chem. Rev. *69*, 157 (1969).

276) Mitchell, K. A. R.: Inorg. Chem. 9, 1960 (1970).
277) Bartell, L. S., Su, L. S., Yow, H.: Inorg. Chem. 9, 1903 (1970).
278) Chandler, G. S., Thirunamachandran, T.: J. Chem. Phys. 49, 3640 (1968).
279) Boyd, D. R.: J. Chem. Phys. 52, 4846 (1970).
280) Lucken, E. A. C.: Struct. Bonding 6, 1 (1969).
281) Heilweil, I. J.: Am. Chem. Soc., Div. Petrol. Chem., Preprints 10, 19 (1965).
282) Dakternieks, D. R., Graddon, D. P.: Australian J. Chem. 23, 1989 (1970).
283) Giancotti, V., Ripamonti, A.: J. Chem. Soc. A 1969, 706.
284) Cavell, R. G., Byers, W., Day, E. D., Watkins, P. M.: Inorg. Chem. 11, 1598 (1972).
285) Sanderson, R. T.: Chemical periodicity. Chap. 3, pp. 37—55. New York: Reinhold Publishing Co. 1960.
286) Sanderson, R. T.: Inorganic chemistry. Chap. 6, pp. 69—88. New York: Reinhold Publishing Co. 1967.
287) Sanderson, R. T.: Chemical bonds and bond energy. Chap. 2, pp. 13—26. New York: Academic Press 1971.
288) A computer program for Sanderson charge distribution calculations is available from H. J. Stoklosa upon request.
289) Stoklosa, H. J.: Unpublished results; — Montgomery, H. E., Stoklosa, H. J., Wasson, J. R.: Unpublished results.
290) Kabachnik, M. I., Mastrukova, T. A., Shipov, A. E., Melentyeva, T. A.: Tetrahedron 9, 10 (1960).
291) Toropova, V. F., Saikina, M. K., Aleshov, R. S.: Zh. Obshch. Khim. 37, 725 (1967); — Chem. Abstr. 67, 26,388r (1967).
292) Toropova, V. F., Cherkasov, R. A., Savel'eva, N. I., Pudovik, A. N.: Zh. Obshch. Khim. 40, 1043 (1970); — Chem. Abstr. 73, 92.187z (1970).
293) Burger, K., Papp-Molnar, E., Vasarhelyi-Nagy, H., Korecz, L.: Magy. Kem. Folyoirat 76, 138 (1970); — Chem. Abstr. 73, 19,124x (1970).
294) Busev, A. I., Byr'ko, V. M.: Tr. Komis. Analit. Khim., Akad. Nauk SSSR, Inst. Geokhim. i Analit. Khim. 9, 59 (1958); — Chem. Abstr. 53, 3840d (1959).
295) Shetty, P. S., Fernando, Q.: J. Inorg. Nucl. Chem. 28, 2873 (1966).
296) Shetty, P. S., Fernando, Q.: Brit. 796,181, June 4, 1958; Chem. Abstr. 54, 1298 (1960).
297) Jackson, B. E.: New Zealand J. Sci. 8, 368 (1965); — Chem. Abstr. 63, 17,754 (1965).
298) Golding, R. M., Jackson, B. E.: New Zealand J. Sci. 8, 383 (1965); — Chem. Abstr. 63, 17,754d (1965).
299) Burn, A. J.: Advan. Chem. Ser. 75, 323 (1968).
300) Kendall, P. F., Rimmer, A.: Chem. Ind. (London) 1962, 1864 and references therein.
301) Hoerding, D., Fischer, H.: Schmierstoffe Schmierungstech. 31, 25 (1968); — Chem. Abstr. 71, 93,279b (1969).
302) Perry, S. G.: J. Gas Chromatog. 1964, 93.
303) Hassan, B. E. M.: Chem. Tech. (Leipzig) 23, 540 (1971); — Chem. Abstr. 75, 117,829s (1971).
304) Hanneman, W. W., Porter, R. S.: J. Org. Chem. 29, 2996 (1964).
305) Legate, C. E., Burnham, H. D.: Anal. Chem. 32, 1042 (1960).
306) Cooks, R. G., Gerrard, A. F.: J. Chem. Soc. B 1968, 1327 and references therein.
307) Damico, J. N.: J. Assoc. Offic. Anal. Chem. 49, 1027 (1966).
308) Ivanyutin, M. I., Busev, A. I.: Nauchn. Dokl. Vysshei Shkoly, Khim. i Khim. Tekhnol. 1958, 73; — Chem. Abstr. 53, 968a (1959).
309) Ivanyutin, M. I.: Chem. Abstr. 64, 8918d (1966).
310) Busev, A. I., Shishkov, A. N.: Zh. Anal. Khim. 23, 1675 (1968); — Chem. Abstr. 70, 34016v (1969).
311) Busev, A. I., Shishkov, A. N.: Zh. Anal. Khim. 23, 181 (1968); — Chem. Abstr. 68, 101,523g (1968).

J. R. Wasson, G. M. Woltermann and H. J. Stoklosa

312) Abdusalyamov, N., Ganiev, A. G., Yuldasheva, K.: Uzbeksk Khim. Zh. *12*, 20 (1968); – Chem. Abstr. *69*, 54,773x (1968).
313) Forster, W. A., Brazenall, P., Bridge, J.: Analyst *86*, 407 (1961).
314) Busev, A. I., Ivanyutin, M. I.: Vestn. Mosk. Univ. Ser. Mat., Mekhan., Astron., Fiz. i. Khim. *13*, 177 (1958); – Chem. Abstr. *53*, 6895b (1959).
315) Ivanov, S. K.: C. R. Acad. Bulg. Sci. *20*, 1153 (1967); – Chem. Abstr. *68*, 95,153t (1968).
316) Masero, M., Perini, M.: Chim. Ind. (Milan) *37*, 945 (1955); – Chem. Abstr. *50*, 4723a (1956).
317) Ripan, R., Eger, I., Mirel, C.: Acad. Rep. Populare Romine, Filiala Cluj, Studii Cercetari Chim. *14*, 49 (1963); – Chem. Abstr. *62*, 3396b (1965).
318) Ripan, R., Eger, I., Mirel, C.: Acad. Rep. Populare Romine, Filiala Cluj, Studii Cercetari Chim. *14*, 57 (1963); – Chem. Abstr. *62*, 3396d (1965).
319) Handley, T. H., Dean, J. A.: Anal. Chem. *34*, 1312 (1962).
320) Zucal, R. H., Dean, J. A., Handley, T. H.: Anal. Chem. *35*, 988 (1963).
321) Larionov, S. V., Shul'man, V. M., Podol'skaya, L. A.: Russ. J. Inorg. Chem. *12*, 1295 (1967).
322) Cabrera, C. A., Woltermann, G. M., Wasson, J. R.: Tetrahedron Letters *1971*, 4485.
323) Angus, J. R., Woltermann, G. M., Vincent, W. R., Wasson, J. R.: J. Inorg. Nucl. Chem. *33*, 3041 (1971).
324) Angus, J. R., Woltermann, G. M., Wasson, J. R.: Inorg. Nucl. Chem. Letters *7*, 837 (1971).
325) Sgamellotti, A., Porta, P., Cervone, E.: Ric. Sci. *38*, 1223 (1968).
326) Dhingra, M. M., Govil, G., Kanekar, C. R.: Chem. Phys. Letters *10*, 86 (1971).
327) Sacconi, L.: J. Chem. Soc. A *1970*, 248.
328) Aoki, T., Yamasaki, T.: Nippon Kagaku Zasshi *85*, 757 (1964); – Chem. Abstr. *63*, 14,230a (1965).
329) Ripan, R., Mirel, C.: Rev. Roumaine Chim. *9*, 567 (1964).
330) Furlani, C.: Coord. Chem. Rev. *3*, 141 (1968).
331) Beech, G., Mortimer, C. T., Tyler, E. G.: J. Chem. Soc. A *1967*, 1111.
332) Busev, A. I., Ivanyutin, M. I.: Tr. Komis. Analit. Khim., Akad. Nauk SSSR. *11*, 172 (1960).
333) Looney, Q., Douglas, B. E.: Inorg. Chem. *9*, 1955 (1970).
334) Kanekar, C. R., Dhingra, M. M., Marathe, V. R., Nagarajan, R.: J. Chem. Phys. *46*, 2009 (1967).
335) Martin, R. L., White, A. H.: Nature *223*, 394 (1969); – White, A. H.: Ph. D. Thesis. University of Melbourne 1966.
336) Sokol'skii, D. V., Kurashvili, L. M., Bersuker, I. B., Budnikov, S. S., Zavorokhina, I. A.: Dokl. Akad. Nauk SSSR *198*, 126 (1971); – Chem. Abstr. *75*, 55,604d (1971).
337) Mitchell, W. J., DeArmond, M. K.: J. Mol. Spectry. *41*, 33 (1972).
338) Hauser, P. J., Schreiner, A. F., Gunter, J. D., Mitchell, W. J., DeArmond, M. K.: Theoret. Chim. Acta *24*, 78 (1972).
339) DeArmond, M. K., Mitchell, W. J.: Inorg. Chem. *11*, 181 (1972).
340) Schmidtke, H.-H., Garthoff, D.: J. Am. Chem. Soc. *89*, 1317 (1967).
341) Cavell, R. G., Day, E. D., Byers, W., Watkins, P. M.: Inorg. Chem. *11*, 1759 (1972).
342) Wasson, J. R.: Nephelauxetic parameters, effective metal charges and spectral intensities of octrahedral chromium(III) complexes. Submitted for publication.
343) Ballhausen, C. J., Gray, H. B.: Inorg. Chem. *1*, 111 (1962).
344) Wasson, J. R., Stoklosa, H. J.: The electronic spectra of high symmetry oxovanadium(IV) complexes – A crystal field approach. Submitted for publication.
345) Blake, A. B., Cotton, F. A., Wood, J. S.: J. Am. Chem. Soc. *86*, 3024 (1964).

346) Ballhausen, C. J.: Progr. Inorg. Chem. 2, 251 (1960).
347) Jørgensen, C. K.: Helv. Chim. Acta, Fasciculus Extraordinarius Alfred Werner 1967, 131.
348) Shupack, S. I., Billig, E., Clark, R. J. H., Williams, R., Gray, H. B.: J. Am. Chem. Soc. 86, 4594 (1964); – Latham, A. R., Hascall, V. C., Gray, H. B.: Inorg. Chem. 7, 788 (1965).
349) Müller, A., Diemann, E.: Chem. Commun. 1971, 65.
350) Siimann, O., Fresco, J.: Am. Chem. Soc. 92, 2652 (1970).
351) Dingle, R.: Inorg. Chem. 10, 1141 (1971).
352) Jensen, K. A., Krishnan, V., Jørgensen, C. K.: Acta Chem. Scand. 24, 743 (1970).
353) Kato, H.: Bull. Chem. Soc. Japan 45, 1281 (1972).
354) Keijzers, C. P., deVries, H. J. M., van der Avoird, A.: Inorg. Chem. 11, 1338 (1972).
355) Il'ina, L. A., Zemlyanskii, N. I., Larionov, S. V., Chernaya, N. M.: Izv. Akad. Nauk. SSSR, Ser. Khim. 1969, 198; – Chem. Abstr. 70, 102,609t (1969).
356) Larionov, S. V., Il'ina: Zh. Obshch. Khim. 41, 762 (1971); – Chem. Abstr. 75, 68,066e (1971).
357) Zemlyanskii, N. I., Gorak, R. D.: Zh. Obsch. Khim. 41, 1691 (1971); – Chem. Abstr. 76, 3497x (1972).
358) Corbridge, D. E. C.: Topics Phosphorus Chem. 6, 235 (1969).
359) Popov, E. M., Kabachnik, M. I., Mayants, L. S.: Russ. Chem. Rev. 30, 362 (1961).
360) Adams, D. M., Cornell, J. B.: J. Chem. Soc. A 1968, 1299.
361) Chaston, S., Livingstone, S. E., Lockyer, T. N., Pickles, V. A., Shannon, J. S.: Australian J. Chem. 18, 673 (1965).
362) Shagidullin, R. R., Lipatova, I. P.: Izv. Akad. Nauk. SSSR, Ser. Khim. 1971, 1024; – Chem. Abstr. 75, 82,002n (1971).
363) Shimazu, A., Ogawa, T.: Junkatsu 11, 195 (1966); – Chem. Abstr. 69, 24,448c (1968).
364) Zimina, K. I., Kotova, G. G., Sanin, P. I., Sher, V. V., Kuz'mina, G. N.: Neftekhimiya 5, 629 (1965); – Chem. Abstr. 64, 162h (1966).
365) Zimina, K. I., Kotova, G. G., Sher. V. V., Kuz'mina, G. N., Sanin, P. I.: Khim. i Tekhnol. Topliv. i Masel 11, 54 (1966); – Chem. Abstr. 64. 10,586d (1966).
366) Zimina, K. I., Kotova, G. G., Sanin, P. I., Sher, V. V.: Tr. Vses. Nauch.-Tekh. Soveshch. Prisadkam Miner. Maslam 1966, 109; – Chem. Abstr. 67, 55,893g (1967).
367) Zemlyanskii, N. I., Turkevich, V. V., Murav'ev, I. V., Chernaya, N. M.: Spektrosk. At. Mol. 1969, 354; – Chem. Abstr. 1970, 119,822s.
368) Watt, G. W., McCormick, B. J.: Spectrochim. Acta 21, 753 (1965).
369) Siimann, O., Fresco, J.: J. Chem. Phys. 54, 734,740 (1971).
370) Siimann, O., Fresco, J.: Inorg. Chem. 8, 1846 (1969).
371) Husebye, S.: Acta Chem. Scand. 19, 774 (1965).
372) Mikulski, C. M., Karayannis, N. M., Pytlewski, L. L.: Inorg. Nucl. Chem. Letters 7, 785 (1971).
373) Martin, R. L., White, A. H.: Transition Metal Chem. 4, 113 (1968).
374) Blomquist, J., Roos, B., Sundbom, M.: Chem. Phys. Letters 9, 160 (1971).
375) Katritzky, A. R., Nesbit, M. R., Michalski, J., Tulimowski, Z., Zwierzak, A.: J. Chem. Soc. B 1970, 140.
376) Grabenstetter, R. J., Quimby, O. T., Flautt, T. J.: J. Phys. Chem. 71, 4194 (1967).
377) Keiter, R. L., Verkade, J. G.: Inorg. Chem. 8, 2115 (1969); – Pidcock, A., Richards, R. E., Venanzi, L. M.: J. Chem. Soc. A 1966, 1707; – Mather, G. G., Pidcock, A.: J. Chem. Soc. 1970, 1226.
378) Eaton, D. R., Phillips, W. D.: Advan. Magn. Resonance 1, 103 (1965).
379) deBoer, E., van Willigen, H.: Progr. NMR Spectry 2, 111 (1967).

380) Webb, G. A.: Ann. Repts. NMR Spectry 3, 211 (1970).
381) Schwarzhans, K. E.: Angew. Chem. Intern. Ed. 9, 946 (1970).
382) Happe, J. A., Ward, R. L.: J. Chem. Phys. 39, 1211 (1963).
383) Dhingra, M. M., Govil, G., Kanekar, C. R.: J. Magn. Res. 6, 577 (1972).
384) Sato, M., Yanagita, M., Fujita, Y., Kwan, T.: Bull. Chem. Soc. Japan 44, 1423 (1971).
385) Valitova, F. G., Ryzhmanov, Y. M.: Dokl. Akad. Nauk. SSSR 170, 1124 (1966).
386) Garif'yanov, N. S., Kozyrev, B. M., Gainullin, I. F.: Zh. Strukt. Khim. 9, 529 (1968).
387) Cowsik, R., Kumari, R. G., Srinivasan, R.: Proc. Nucl. Phys. Solid State Phys. Symp., 14th, 1963, pp. 175–8; – Chem. Astr. 75, 114,573t (1971).
388) Henrici-Olive, G., Olive, S.: Angew. Chem. Intern. Ed. 9, 957 (1970).
389) Henrici-Olive, G., Olive, S.: Chem. Commun. 1969, 596.
390) McCormick, B. J., Featherstone, J. L., Stoklosa, H. J., Wasson, J. R.: Oxovanadium(IV) complexes of dimethyldithioarsinate. Inorg. Chem., in press.
391) Zverev, G. M., Prokorov, A. M.: Soviet Phys. JETP 7, 707 (1958).
392) Kenworthy, J. G., Myatt, J., Todd, P. F.: Chem. Commun. 1969, 263.
393) Gregorio, S., Weber, J., Lacroix, R.: Helv. Phys. Acta 38, 172 (1965); – Gregorio, S., Lacroix, R.: Proc. Colloq. AMPERE 12, 213 (1963).
394) Garif'yanov, N. S., Kozyrev, B. M., Luchkina, S. A.: J. Struct. Chem. 9, 794 (1968); – Zh. Strukt. Khim. 9, 901 (1968).
395) Orton, J. W.: Electron paramagnetic resonance, pp. 113 and 140. London: Sliffe Books Ltd. 1968.
396) Henning, J. C. M., Bongers, P. F., van den Boom, H., Voermans, A. B.: Phys. Lett. 30A, 307 (1969).
397) Garif'yanov, N. S., Kamenev, S. E., Ovchinnikov, I. V.: Teor. i Eksperim. Khim. 3, 661 (1967).
398) Garif'yanov, N. S., Troitskaya, A. D., Razumov, A. I., Gurevich, T. A., Kondrat'eva, O. I.: Russ. J. Inorg. Chem. 16, 563 (1971).
399) Dalton, L. A., Dalton, L. R., Brubaker, C. H.: Progress in coordination chemistry (ed. M. Cais), p. 526. New York: Elsevier Publishing Co. 1968.
400) Garif'yanov, N. S.: Zh. Strukt. Khim. 12, 170 (1971).
401) Garif'yanov, N. S., Kamenev, S. E., Kozyrev, B. M., Ovchinnikov, I. V.: Dokl. Akad. Nauk. SSSR 177, 880 (1967).
402) Brackett, G. C., Richards, P. L., Wickman, H. H.: Chem. Phys. Letters 6, 75 (1970).
403) Garif'yanov, N. S., Luchkina, S. A.: Izv. Akad. Nauk SSSR, Ser. Khim. 1969, 471.
404) McDonald, C. C., Phillips, W. D., Mower, H. F.: J. Am. Chem. Soc. 87, 3319 (1965); – Crow, J. P., Cullen, W. R., Herring, F. G., Sams, J. R., Tapping, R. L.: Inorg. Chem. 10, 1616 (1971).
405) Beckett, R., Heath, G. A., Hoskins, B. F., Kelly, B. P., Martin, R. L., Roos, I. A. G., Weickhardt, P. L.: Inorg. Nucl. Chem. Letters 6, 257 (1970).
406) Golding, R. M., Sinn, E., Tennant, W. C.: J. Chem. Phys. 56, 5296 (1972).
407) Zamaraev, K. I.: Zh. Strukt. Khim. 10, 32 (1969); – Chem. Abstr. 70, 110,424u (1969).
408) Solozhenkin, P. M., Kopitsya, N. I.: Izv. Akad. Nauk Tadzh. SSR, Otd. Fiz.-Mat. Geol.-Khim. Nauk 1968, 56; – Chem. Abstr. 70, 15,871d (1969).
409) Garif'yanov, N. S., Kozyrev, B. M.: J. Struct. Chem. 6, 734 (1965); – Zh. Strukt. Khim. 6, 773 (1965).
410) Solozhenkin, P. M., Kopitsya, N. I., Grishina, O. N.: Zh. Strukt. Khim. 12, 167 (1971).
411) Larin, G. M., Solozhenkin, P. M., Kopitsya, N. I.: Izv. Akad. Nauk, Ser. Khim. 1969, 475.
412) Larin, G. M., Solozhenkin, P. M., Dyatkina, M. E., Kopitsya, N. I.: Zh. Strukt. Khim. 12, 26 (1971).

413) Solozhenkin, P. M., Kopitsya, N. I., Loseva, N. P.: Theor. i Eksperim. Khim. *4*, 708 (1968).
414) Vashman, A. A.: Russ. J. Inorg. Chem. *12*, 1509 (1967).
415) Shopov, D., Yordanov, N. D.: Inorg. Chem. *9*, 1943 (1970).
416) Yordanov, N. D., Shopov, D.: Compt. Rend. Acad. Bulgare Sci. *23*, 1239 (1970).
417) Yordanov, N. D., Shopov, D.: Inorg. Chim. Acta *5*, 679 (1971).
418) Vezentan, N.: Studia Univ. Babes-Bolyai, Ser. I Math., Phys. *14*, 91 (1969); – Chem. Abstr. *71*, 65,887e (1969).
419) Zelonka, R. A., Baird, M. C.: Chem. Commun. *1971*, 780; – Can. J. Chem. *50*, 1269 (1972).
420) Theriot, L. J., Ganguli, K. K., Kavarnos, S., Bernal, I.: J. Inorg. Nucl. Chem. *31*, 3133 (1969).
421) Rehorek, D., Kirmse, R., Thomas, P.: Z. Anorg. Allgem. Chem. *385*, 299 (1971).
422) van Rens, J. G. M., Keijzers, C. P., van Willigen, H.: J. Chem. Phys. *52*, 2858 (1970).
423) Gregson, A. K., Mitra, S.: J. Chem. Phys. *49*, 3696 (1968).
424) Kirmse, R., Wartewig, S., Windsch, W., Hoyer, E.: J. Chem. Phys. *56*, 5273 (1972); – Kirmse, R., Windsch, W., Hoyer, E.: Chem. Phys. Letters *4*, 565 (1970).
425) Muller, A., Rao, V. V. K., Christophliemk, P.: J. Inorg. Nucl. Chem. *34*, 345 (1972) and references therein.
426) Alison, J. M. C., Stephenson, T. A.: Chem. Commun. *1970*, 1092 and references therein.
427) Grim, S. O., Matienzo, L. J., Swartz, W. E.: J. Am. Chem. Soc. *94*, 5116 (1972).
428) Sadovskii, A. P., Larionov, S. V., Vainshtein, E. E.: Zh. Strukt. Khim. *8*, 1043 (1967); – Merritt, J., Agazzi, E. J.: Anal. Chem. *38*, 1954 (1966); – Nefedov, V. I., Mazalov, L. N., Sadovskii, A. P., Bertenev, V. M., Porai-Koshits. M. A.: Zh. Strukt. Khim. *12*, 1015 (1971); – Nefedov, V. I., Narbutt, K. N.: Zh. Strukt. Khim. *12*, 1069 (1971).

Received August 21, 1972.

Topics in Current Chemistry
Fortschritte der chemischen Forschung

Preisänderungen vorbehalten
Prices are subject to change without notice

W. Kutzelnigg, G. Del Re, G. Berthier

σ and π Electrons in Organic Compounds

11 figures
IV, 122 pages
1971
Soft cover DM 48,—

Fortschritte
der chemischen
Forschung/
Topics in
Current Chemistry,
Vol. 22

In the development of quantum chemistry few concepts have proved to be as significant as the distinction between σ and π electrons in organic compounds. This distinction suggested the approximation known as the ,σ-π separation' which has made it possible to calculate many important physical and chemical properties of unsaturated compounds within the frame of a ,pure π-electron theory'. This type of theory, which goes back essentially to E. Hückel, has the advantage of great conceptual and practical simplicity and has been successful in solving many problems. Nowadays, the advent of computers has made feasible treatments of polyatomic molecules of small and medium size, taking into account all the electrons. Nevertheless, scientific economy suggests that, if certain physical or chemical facts can be understood in terms of π electrons only, one should try to do so; therefore, 'π-electron theories' still deserve analysis and applications.
The authors begin by restating and clarifying the basic concepts on which the whole question of the σ—π separation rests. They consider the conditions under which the electrons of a molecule can be classified into σ and π electrons and indicate what should be understood by 'σ—π separation' and what are the limitations of this approximation. They show that the most important part of the 'σ—π interaction' is usually taken into account within the σ—π separation scheme and finally discuss whether the σ—π interaction has a significant effect on the theoretical predictions made for the physical properties of unsaturated molecules. To preserve the rigour of certain arguments, a number of quantum-mechanical formulas are useful; they are introduced, when required, with the necessary explanations on notations. (148 references)

Springer - Verlag
Berlin Heidelberg New York
London München Paris Sydney Tokyo Wien

In kritischen Übersichten werden in dieser Reihe Stand und Entwicklung aktueller chemischer Forschungsgebiete beschrieben. Sie wendet sich an alle Chemiker in Forschung und Industrie, die am Fortschritt ihrer Wissenschaft teilhaben wollen. In der Regel werden nur Beiträge veröffentlicht, die ausdrücklich angefordert worden sind. Schriftleitung und Herausgeber sind aber für ergänzende Anregungen und Hinweise jederzeit dankbar. Manuskripte können in den „Fortschritten der chemichen Forschung" in Deutsch oder Englisch veröffentlicht werden.

Jeder Band der Reihe ist einzeln käuflich.

This series presents critical reviews of the present position and future trends in modern chemical research. It is addressed to all research and industrial chemists who wish to keep abreast of advances in their subject.

As a rule, contributions are specially commissioned. The editors and publishers will, however, always be pleased to receive suggestions and supplementary information. Papers are accepted for "Topics in Current Chemistry" in either German or English.

Any volume of the series may be purchased separately.

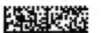